DAM BREAK
IN GEORGIA

Sadness and Joy
At Toccoa Falls

DAM BREAK
IN GEORGIA

Sadness and Joy
At Toccoa Falls

by
K. NEILL FOSTER
with
Eric Mills

HORIZON HOUSE PUBLISHERS
3825 Hartzdale Drive
Camp Hill, Pennsylvania 17011

ISBN 0-88965-023-3

HORIZON BOOKS
3825 Hartzdale Drive
Camp Hill, Pennsylvania 17011

Printed in United States of America

to our bereaved brothers and sisters
who were willing to open
still-fresh wounds
to tell this gripping story
to us — and to the world

Contents

Foreword

This is an appealing book, a clear presentation of people who lived and died with extraordinary strength. It will help you face life and death with optimism.

It is the record of a disaster with a difference. It catapults the small north Georgia city of Toccoa into the awareness of the world. It tells what happened to a college community when 176 million gallons of water smashed through the campus in the early hours of November 6, 1977.

You are about to meet unique people. From all outward appearances they are very ordinary. But looks can be deceiving. Some of these people you will never forget. I knew most of the people who died and all of those who survived. Knowing them has enriched my life. And your life may change too as you read this wonderful account of sadness and joy at Toccoa Falls.

No pretense is made of telling the whole story of horror and subsequent victory. Several books could be written about the valor and compassion shown in this small community. Those books will have to wait for another time. But I commend **Dam Break in Georgia** to you as the beginning of a powerful story which is still unfolding.

There are thousands who have stepped in to help Toccoa

Falls College since the disaster and we are deeply grateful. Among the most heartfelt and spontaneous responses have been those which have come from the public officials, businesses, organizations and citizens of Toccoa, Georgia. To these, especially, we offer our sincerest thanks.

Kenn W. Opperman
Toccoa Falls College
Toccoa Falls, Georgia

A Letter
of
Introduction

March 14, 1978

Dear Friends,

This is a story about faith. It is a story that will affect
every person who has ever doubted, for this is a personal
testimony that there is inherent courage within us to face
the challenges of life and death.

I visited Toccoa Falls College on the day after the disaster
that you will read about in this book. I went because I hoped
that I could comfort those who had survived. Instead I was
enveloped by hope and courage and love.

The miracle of Toccoa Falls confirms what I believe. God
loves us and will help us always. He gives us unlimited
strength when we trust in him.

This is a story that will never have an ending.

Sincerely,

Rosalynn Carter

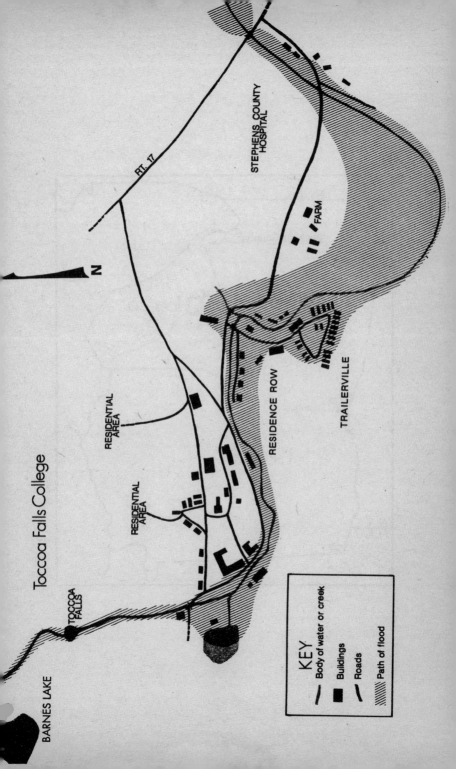

Toccoa Falls College

BARNES LAKE

TOCCOA FALLS

RT. 17

N

RESIDENTIAL AREA

RESIDENTIAL AREA

STEPHENS COUNTY HOSPITAL

FARM

RESIDENCE ROW

TRAILERVILLE

KEY
Body of water or creek
Buildings
Roads
Path of flood

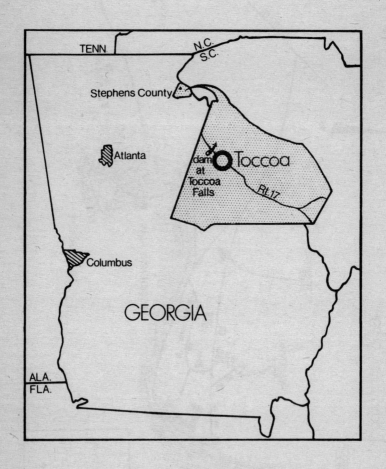

I.
Flood over the Falls

Chapter 1

Daddy, Don't Let Me Go!

Barbara Eby literally leaped out of bed just before the water came.

"The dam's broken!" she shouted, not even knowing how she knew.

Dave Eby, her husband and the Dean of Men at Toccoa Falls College, was sleepy and incredulous. "You're crazy!" he said. "Get back in bed." Then he heard sounds that brought him instantly to a full alert.

Barbara hurried to the bedroom where their three small children, Kimberly, Kevin, and Kelly, were asleep. As she headed down the hall she noticed that the night light was still on in the bathroom. Almost simultaneously, she saw it go off and heard the rushing water.

Just that week she had moved all three children into one room to make available another room in which they could play. She headed first for Kimberly, 7, and carried her to where Dave was in the master bedroom.

As Barbara headed back for Kevin, 5, and Kelly, 3, Dave tried to bash out a window in the bedroom. As he smashed the glass and the frame, blood poured from his arm. But he still wasn't through to the outside. Beyond

the glass there was plastic which he could not penetrate—
the water was already up to that level. Punching the
plastic, Dave says, was like punching Jell-O.

Suddenly a torrent of water burst through the plastic.
The powerful current swept Dave and Kimberly out of
the bedroom and slammed them into the utility room.
Eventually, it washed them right out of the house.

Meanwhile, Barbara had troubles of her own.

When she reached the children's bedroom she swept the
other two children into her arms. One in each arm, she
stepped into the hallway. Just as she did so, more water
poured in and she heard Dave shout, "Hold on, we're
going!"

The water was coming rapidly now, but she braced herself in the hallway and managed to keep her footing. She
waded through the water to a room divider just past the
end of the corridor. By the time she reached it the water
was to her chest.

She could hear Dave and Kimberly struggling in the
utility room and wanted to join them. But then the re-
frigerator tipped, blocking the door.

Barbara knew she and the children had to stay above the
water. The divider, a solid structure about chest-high,
seemed to be the answer. And reaching up from that
divider to the ceiling were three decorative wooden
spindles.

She told the children to hold on to her tightly while she
attempted to climb onto the divider's level counter. As
she struggled she got all wrapped up in her long night-
gown, but she somehow got her knee on the counter and
rose to her feet, then climbed part way up one spindle,
the children desperately clinging to her all the while.

From there she shouted, "David! David!" But there was
no answer. She told Kevin that Daddy and Kimberly were
dead, and that just the three of them were left. Her young
son began to pray earnestly.

Household furniture was banging around Barbara and
the children now, but the divider served to shield them

from it. Then the spindles ripped loose from the ceiling, fell over about a foot, and little Kelly almost slipped from Barbara's grasp into the swirling water.

But somehow Barbara held on to her, though for how long she could continue to do so, she didn't know. Her arm felt weak now and numb, and she told Kevin, who was at the top of the spindle, that he would simply have to hang on for himself. If she was going to keep Kelly's head from going underwater she would need both her hands.

Though he had been washed from the house, Dave Eby had somehow managed to retain his hold on Kimberly. He recalls feeling shingles in the water outside and clambering onto the roof of his home. Kimberly was still holding on.

Dave says the sound on the rooftop was like the roar of a jet taking off. Sitting there on the peak, both father and daughter felt sure that the rest of the family had been lost.

Dave was worried about the trees that the flood was tossing around like toothpicks. Some were two feet in diameter and he feared that one of them would topple onto the roof of the house.

From his vantage point David saw a partition about five feet wide floating near the creek bank — which was only about ten feet from the house — so he decided to put his little girl on his back and jump. Had he been a swimmer he might have jumped beside the partition instead of on it. But when his feet struck the partition it shattered and little Kimberly lost her grip on her daddy and disappeared into the current. Dave went down, too.

Dave recalls, "I thought it was the end, and I knew if I took another breath I'd be in heaven. But I thought I'd better not — it could be suicide! Coming back up I bumped into the house, and then I felt two small arms wrap themselves around my neck and heard my little girl giggle, 'Daddy, don't let me go again!' She was laughing."

21

Dave had surfaced by the front door of his battered home. From there, with his head above the water and Kimberly on his back, he could shout to his wife and the children inside.

He grabbed the frame of the screen door and pulled himself inside. Straddling the screen door, with the water up to his chest, he heard his attractive young wife say from the spindle, "I'm too tired, I can't make it."

They prayed earnestly then. Dave says he shouted at God — he really wanted Him to hear!

The water continued to rise. Then Kevin prayed too. Clinging to a spindle for dear life, he asked the Lord to lower that water.

Incredibly, and much to their relief, the water fell about two feet.

The crisis had passed. The family was safe. Barbara inched down to where she could touch the floor once again. Then she lowered the children to the solid counter. David looked for a way out of the house to safety.

Soon, over destruction and debris, the Eby family clambered up the rocky cliff against which their floating house had lodged. Since they all had survived they could not believe that anyone else had died.

The dam had burst. The worst natural disaster in the State of Georgia since 1939 had already hungrily devoured little children and devastated sleeping families.

Nearly six percent of the college community had been wiped out in one cruel stroke. Most of the damage, at any given place, occurred in about twenty seconds.

November 6, 1977, was barely an hour and thirty minutes old.

The story does not begin, of course, with the sounds of the Toccoa Falls flood. The roar of several freight trains, fierce and terrible in the narrow valley, was but one description of the thunderous noise surrounding a sad

22

drama that became more painful — and triumphant — as events unfolded.

> *The voice of the Lord is upon many waters: the God of glory thundereth: the Lord is upon many waters.*
>
> Psalm 29:3

Chapter 2

He Fingered His Ten Dollar Bill

"I surely appreciate your kindness, sir. And the price is certainly reasonable. Only..."

"Only what?" barked the older man.

"It's about the terms, sir." Mr. Forrest reached into his pocket and fingered his ten dollar bill — with which he hoped to found a college at Toccoa Falls, Georgia.

"What about the terms? I haven't even mentioned them yet."

"I know — but I have only ten dollars, and no tangible place to get any more."

Today, Toccoa Falls College is situated just outside the city of Toccoa, 100 miles northeast of Atlanta and seven miles from the border of South Carolina.

Situated on 1,100 acres of choice property between the Great Smoky Mountains and the Piedmont, Toccoa Falls College is strikingly beautiful. On college-owned property is located Toccoa Falls, a 186-foot cataract that plunges into a gorge-like box canyon. Toccoa Creek then meanders through the campus, under the concrete bridge at Georgia's Route 17, and eventually empties into Lake Hartwell, one of the largest man-made lakes in the

United States.

Toccoa Falls College was begun by Dr. R.A. Forrest in a rather dramatic way. Forrest had only $10.00 in his pocket—and nowhere to get more. Undaunted, he approached the banker in Toccoa. The man owned a very desirable piece of property and Forrest knew he would want money, a sizable down payment, and good terms. And he had a right to ask them. But young Forrest wanted something too. So he breathed a silent prayer, wet his lips and stammered, "How much do you want for the hotel?"

"Twenty-five thousand dollars is a fair price, it seems to me. That will include the hotel and about one hundred acres surrounding it."

"Will Toccoa Falls be on that hundred acres?"

"Yes, sir. There's a little piece of land that won't be included though. My power plant is built there, and I'll retain that small piece of property and the water rights to Toccoa Creek above the falls."

"Fair enough, sir," replied Mr. Forrest. "But tell me, what about our lights here?"

"Oh, we'll furnish you lights with no obligation as long as there's water to turn the generator."

At that point, Forrest, with just ten dollars to back him up, made his rather daring offer.

As the businessman drew his breath and opened his mouth, the young minister watched, fascinated yet fearful. Suddenly R.A. Forrest's words tumbled out.

"Sir, I know this must sound presumptuous to you, a businessman and a banker. But I know God wants us to establish a school for Him here. Sir, I'll pay you ten dollars (I have only that and my return ticket to Atlanta), and the Lord and I will owe you the other $24,990.00 on the $25,000.00 deal—if you'll trust the Lord and us."

"I can trust the Lord," came the dry response. The businessman held out his hand for the ten dollar bill.

And so it was that the beautiful property on a placid

creek was purchased in 1911.[1] It was that same gurgling stream which became, that fateful morning of November 6, 1977, a howling monster of incredible brutality and power.

About one-half mile back from the lip of Toccoa Falls stood Kelly Barnes Dam, a structure built in 1887 of red Georgia earth. Behind the dam was Kelly Barnes Lake, fifty-five acres of placid, fish-filled water. The lake contained 700,000 tons of water, 176 million gallons, and from the air it was one of the most scenic lakes in Georgia. The bottom of the lake, since it was man-made, contained some vegetation and the tall stumps of some trees that had remained over the passing of time.

The Army Corps of Engineers stated that after the dam broke, the waters at the top of Toccoa Falls had developed a terrible velocity. In the half mile between the dam and the lip of the escarpment Toccoa Creek drops 600 feet. The falls itself is 186 feet high.

The water may have come over the falls at 150 miles per hour. It could have had a velocity of 120 miles per hour as it exploded out of the canyon. Certainly, as it slammed into Morrison Hall, an empty dormitory that stood in its path, only solid rock could have stood before it. Hydrologists on the scene would say only that the water could have been going between 50 and 150 miles per hour.

Prior to the disaster, college personnel periodically inspected the dam. In fact, just hours before it broke, two volunteer firemen, Chief David Fledderjohann and his aide Ron Ginther, ascended the slippery road and observed the level of the lake. Water seemed to be emerging into Toccoa Creek in the usual manner. The surface of the lake was several feet below the top of the dam and the secondary spillway for excess water was not in use.

There was, however, some concern about the dam. Observation, even, of its condition — but no apparent cause for alarm.

---

1. *Achieving The Impossible With God,* Moothart, 1956, Toccoa Falls College, Toccoa Falls, Georgia. Used by permission.

Below, directly in the path of the pent-up water, lay a sleeping campus.

As nearly as can be guessed, at approximately 1:25 a.m. the pressure of the water behind Kelly Barnes Dam somehow became irresistible. Unseen by human eye, the powerful water tore at the earthen dike. Tiredly — tragically — it gave up.

And suddenly all the fury of death itself was surging down the slope toward the lip of Toccoa Falls.

Dave Eby, Dean of Men at the college, says he sensed two distinct things — first, a vibration, then a thud. This may have been the first rush of boulders and debris over the falls. A student heard a sound like the puncturing of a balloon.

The water plunged over Toccoa Falls, a cataract thirty feet higher than Niagara. In those awful moments the morning of November 6, it really was Niagara.

There was a turbine-like effect. The falls and box canyon created a perfect funnel and, for split seconds at least, the canyon held all the pent-up violence. But almost instantly the gorge was full and at its narrow mouth the water spewed out in all its horror and fury. The action was similar to that of a fire hose. The mud and uprooted trees, falling boulders and roaring water — all were vomited out of the canyon.

As the waters boiled out of the gorge a power pole bearing a transformer toppled and electric lines were tossed into the river where they snarled and snapped like high-voltage whips. The effect was such that one of the men staying in Forrest Hall, the men's dormitory, compared it to the Fourth of July.

Massive geysers surged skyward as the water slammed into obstacles in the narrow valley. Debris was tossed as high as sixty-five feet above the surface of the creek under normal conditions. Cars were twirling and smashing into each other.

Another immediate result was that for the second time that evening, the lights went out on the college campus.

27

They also went out in the residences in the path of the flood and in Trailerville, a cluster of mobile homes occupied by married students and their families. The darkness compounded the fury of the flood.

Gate Cottage, the first college building to feel the force of the waters, stood just west of the canyon's mouth. Curiously, it still stands. Its stone-front walls may help explain why, but the real reason for its survival is that it stood just out of the way of the main rush of water. Even so, the current collapsed doors and windows and bumped pieces of furniture around before flowing on through. Strangely, some of the glasses in the restaurant at Gate Cottage remained on their shelves, some of the dishes stayed intact in their cabinets. And a roll of paper towels was wet only part way through. Had the dam burst eleven hours later, perhaps sixty people would have been dining there in the restaurant, one of the finest in the area.

In seconds the water reached the Bandy residence. Dr. Julian Bandy is a former President of the college (seventeen years) and has also served nine years as Vice President of the Christian and Missionary Alliance, the religious society with which the independent college is associated. Bandy's administration was one of development, consolidation and building, and it was under his leadership that many of the structures soon to disappear were built.

Dr. Bandy was off preaching at Naples, Florida at the time, but his 24-year-old grandson, Greg, was still at home. Hearing the roar of the water, he looked out of the window. All he saw was water. Though his window was perhaps thirty feet above the level of the creek, he couldn't even see Forrest Hall, the four-story dormitory across the narrow valley. He was the first human witness of the flood.

Possibly, the darkness hindered observation. Perhaps the power failure made the dormitory invisible. But just possibly the rush of the water out of the canyon and the

28

hose-like effect had actually created a hump of water. If it really was moving at over 100 miles per hour at that point, the water had not yet had time to spread out and do its awful damage. There had been a lot of water in the creek. As later testimony will show, the rise was both gradual and sudden. Apparently there were two major waves—the first about five feet high and the second about thirty feet high.

The boxing effect of the gorge plus the explosive surge out of the canyon turned the full power of the flood away from Gate Cottage, away from the Bandy residence.

It also certainly saved the lives of the David Eby family, next in the path downstream.

Their home was sufficiently to the right of the full force so that when it was hit, it moved off its foundation and slammed against the mountain. And there it came to rest, mercifully spared from the total disintegration it surely would have suffered had it been closer to the creek.

The night before, Barbara Eby's parents had visited them. In their conversation they had asked themselves, "What would it be like if the dam ever broke?" Their discussion, the Ebys say, was not because of any real distrust of the dam, but because of the rising water in the creek. They visualized the damage much as it occurred hours later. But not once did they conceive of the dreadful toll of lives that would be taken.

> *God is our refuge and strength, a very present help in trouble. Therefore will not we fear...though the waters...roar and be troubled....*
>
> **Psalm 46:1-3**

29

Chapter 3

Seven Seconds in Forrest Hall

Across the creek there was also drama in the corridors of Forrest Hall. Kenny Carroll, a lanky lad from Maryland, dashed out of his room and ran the wrong way — toward certain death.

Kenny had arrived home from a date with Marcy Rees late that evening. Because so much water was coming over the falls he wanted to see it, so he walked up the landscaped path into the canyon and gazed at the thundering falls. The time was about twelve midnight. He observed that the landscaped Rose Garden was underwater. Carroll did not know it, but he was in imminent danger as the dam probably had already begun to give way.

After returning to Forrest Hall, and retiring for the night, he was abruptly awakened. He heard sloshing and thought some of the guys were fooling around in the shower again. But when the water began to rush in around his feet he knew differently. As it rose higher and higher he thought, "I'll be dead if I slip."

In the hall he turned the wrong way and ran into fellow students Bobby Carter and Jon Kerr. They turned him around and the trio headed for a door leading into the stairwell. They reached the door about the same time and all three seemed to squeeze through it at once. Carroll estimates that the series of events involved only seven to ten seconds.

Forrest Hall is a four-story brick dormitory with a capacity of 147 students. Normally, when all were present, 124 men were in residence. On the weekend of the flood, however, there were fewer.

Forrest Hall was situated far enough to the east of the water's fury that it was not structurally damaged. But the water rose fast.

Chuck Dowell, a cheerful redhead from Ohio, recalls that in seconds the water penetrated the lower floor. Windows collapsed with the pressure as he lunged toward the door.

Had the windows not popped out as they did the whole first floor could have collapsed. The bearing walls for three more floors of dormitory space would then have been gone. Easily, seventy-five men could have died.

Dowell still does not recall how he got into the corridor. Later, his door was discovered to be jammed solid and had to be broken down.

Chuck's first recollection is reaching for the yellow trash can in the basement hallway. The water was very mucky and the pockets of the cutoffs he was wearing were later found to be filled with slime. Chuck made it to the stairs, clambered up and got outside, then he headed for the girls' dorm to make sure his girlfriend was safe.

Gerry Brittin, Rick Swires and Cary Hanna were not so fortunate. Locked into their rooms by the rising water, they were trapped. Only one body was found in the dormitory. One was found about a half mile away and the other nearly five miles away.

Friends relate that a few days earlier Cary Hanna had talked with them about heaven. Perhaps he was waiting and thinking about the eternal shore. As it turned out, the wait was not long.

31

Rick was planning to be married and the date was set for June 3, 1978. But maybe he too had sensed something, because he had spoken about the possibility of his dying.

Gerry was a pastor's son, from Olean, New York. A dedicated student, he was hard at work on his senior paper and he had turned in early, as was his custom.

The juggernaut howled on, leaving in its wake travail and terror in Forrest Hall. But amazingly, only three died there.

The edge of the flood ripped at the side of the music building. Quickly the side gave way. Thirteen pianos swirled away and doors and furniture were tossed about. The damage was severe, but not complete — because the full force of the flood went elsewhere. And no one was in the music building.

The unrestrained waters headed for Morrison Hall. Once it had been used for cows, and later, when it had become unfit for cows, someone joshed, it had been used for boys! Now, it was directly in the flood path. The wood frame structure was no match for the wall of water, bouncing boulders, and surging mud. Trees turned into battering rams. Morrison Hall was swept away. Fortunately, again, it was an empty building, not in use. No lives were lost.

At the edge of the flood, the waters clawed at the hillside. They nearly reached the graves of Dr. and Mrs. R.A. Forrest, who had begun the college in 1907. In 1911 the school was moved to Toccoa Falls.

Two years later, in 1913, Haddock Inn, where Toccoa Falls College was first housed, had burned to the ground. R.A. Forrest was stunned but hardly ready to quit. As he idly poked a stick through the ashes, God spoke to him with these words from Isaiah 61:3: "...unto them that mourn in Zion, to give unto them beauty for ashes, the oil of joy for mourning, the garment of praise for the spirit of heaviness; that they might be called trees of righteousness, the planting of the Lord."

He believed the thought was from God and went on to rebuild Toccoa Falls College, confident it was not to end in ashes, confident it was indeed "the planting of the Lord." He also became convinced that a certain psalm was very special: "We went through fire and through water, but thou broughtest us out into a wealthy place" (Psalm 66:12).

Forrest believed he had been through his "fire." The development in the intervening sixty-five years has no doubt been beyond his dreams. And he lived until 1957, long enough to see many of his hopes realized.

It appears to have remained for the administration of Dr. Kenn Opperman, who became President in 1974, to go with the college "through water."

Opperman now believes, as Forrest and Bandy believed before him, that the school will certainly emerge, as the psalm implies, into a "wealthy place."

As this account unfolds, it will be seen that unusual events have indeed followed the disaster.

Even if the waters had reached the founders' graves, they could not, of course, have touched Dr. and Mrs. Forrest. They themselves, as they so long had taught and known, were "absent from the body...present with the Lord." But the hungry waters hurtled on — to devour the living.

Next in order, and directly in line, was Ralls Hall, also unoccupied. A frame building with large windows, nothing could save it from becoming kindling, from being added to the now deadly burden of flotsam and debris that was being aimed malevolently at the residences downstream.

Interestingly enough, Dave and Barbara Eby had lived in Ralls Hall when it was used as a high school dormitory. At that time twenty-eight boys had been housed there plus the five members of the Eby family.

But the high school had been closed. Great difficulties had hindered and obstructed the continuation of high school education on the college campus. The decision to

33

close was not uncontested, however, and feelings ran high. One man, not identified, held Dr. Opperman to blame. Moreover, he held a deep resentment against the college President.

After the flood he wrote to ask forgiveness and to apologize. He was sure that had the academy not been closed, his son would have been among the fatalities. Indeed, he might have been. Ralls was in the center of the fury. No one could have survived. The next day not even a trace of the foundation was left.

To the left of the descending torrent sat the guard house, a lightweight wooden structure just big enough for two security guards. Jim Conaway and Paul Tate had been on duty there. Aware that the firemen were monitoring the water levels, they had felt no particular concern.

Then they left to go up to the food service area in the basement of LeTourneau Hall to check the compressors on the freezers. While the guards were inspecting the freezers, their security shack was swept away.

Jim Conaway had heard a very loud noise which he had taken to be wind. He had complained, "That's all we need after all this rain—a big wind!" But when the guards noticed the bushes weren't moving they knew it wasn't wind and quickly stepped outside.

They were anxious now, so they rushed down to look over the wall between Seby Jones Library and Fant Hall. Ordinarily, the gurgling creek lay below.

But Conaway, especially, was stunned by what he saw through the darkness. "There was enough water," he says, "to scare the living daylights out of me."

Where the flimsy shack that they had just vacated moments before had been, there was now a lake.

My times are in thy hand.

Psalm 31:15

34

Chapter 4

The Water Turns Red

Eldon Elsberry, a 36-year-old Nebraskan, splashed across one of the front yards in Residence Row, a series of houses located near the creek. The water was rising, and just as he reached the road connecting the bridge to Trailerville he heard a noise like a shotgun. It was an electric fuse. The 6100-watt line bringing power into the campus had gone out. Seconds later there was another bang — the second fuse had blown. Elsberry and Fire Chief Dave Fledderjohann could not know it, but the transformer had already toppled at Gate Cottage and horror was heading their way.

Elsberry said, "Dave, what would cause the fuses to blow?"

"I don't know," responded Fledderjohann.

But they both felt apprehension.

Suddenly the black water all around them turned red from the Georgia clay.

Earlier, the security guards on campus had been concerned about the rising water in Toccoa Creek. Their concern had focused on the bridge which gave sole access to the residential sections and to Trailerville. It appeared that the bridge might wash out and at one point the water had risen a foot or so above the deck of the bridge. About 12:15 a.m. guard Jim Conaway called Eldon Elsberry. Eldon, in turn, contacted David Fledderjohann.

Earlier still, when Elsberry had returned to the campus, he noticed that the steel gate on the alternate entrance was open. He felt some concern. It was always closed after dark so that all traffic had to pass security.

Near the gymnasium, he encountered Fledderjohann, who assured him that there was no fire, but that the water was rising.

Elsberry went home at that point and changed into fireman's gear. The water over the bridge seemed lower, only eight or nine inches deep now, but still they remembered the creek, which had flooded in the past. Present by that time were some other members of the department: Ron Ginther, Bill Ehrensberger, and Dee Pinney.

About 10:30 p.m., Ron Ginther casually mentioned Kelly Barnes Dam and it was decided to go up and check. Taking the fire department vehicle, David and Ron went up, and visually inspected the road and the dam. They noticed that the water inside the dam was quite far down and apparently in no danger of overflowing. They radioed back by citizen's band that the dam was solid and that there was no need for concern or alarm.

After they had come back, Elsberry went over to Fledderjohann's vehicle and leaned on the door.

"What did you find?" said Elsberry. Fledderjohann replied, "It's as normal as ever. I've seen it much higher many times."

And the water continued to recede until it was once again flowing under the bridge to Trailerville.

Fledderjohann and Elsberry went home. Just when Elsberry was about to retire, the phone rang. It was Jim Conaway at the guard shack.

Conaway was concerned because the bridge seemed unstable. Concluding the telephone conversation, Elsberry said, "I'll talk to David Fledderjohann."

Then, Elsberry phoned David. Neither felt any sense of danger, but they decided to check again.

Bill Ehrensberger also came out. His wife and children were all asleep in the house. In Ron Ginther's trailer, the light was still on. He and his wife invited the rest of the firemen to come in about 1:00 a.m.

After coffee and cookies the firemen left. They noticed that the water seemed to be rising again. The horizontal yellow stripe on Elsberry's boots had been clearly visible when they had sloshed to Ginther's trailer. Now the yellow stripe could not be seen.

One of the firemen, Dee Pinney, lingered and fussed with a ham radio.

Outside, sensing impending danger, Fledderjohann decided to move two families threatened by the rising creek — the Sproulls and the Woerners. Their houses were lowest in the residential areas and Sproulls' house was likely to be flooded if the waters continued to rise.

Fledderjohann went to Sproulls' and Dr. Jerry Sproull came to the door. He hesitated and was reluctant to leave, but finally said, "Let me get some clothes."

At this juncture the fuses blew and the explosions reverberated through the valley. Then abruptly the water turned red — and ominous.

Elsberry wheeled to see a wave of water four or five feet high rolling along soundlessly like rapids. "Look out!" he shouted. "There's a wall of water!"

Fledderjohann shouted back to Bill Ehrensberger and Elsberry, "Try to get back across the creek and sound the alarm."

Ehrensberger made it to their truck — on the driver's side. Elsberry ran through the deepening water to the truck only scant feet away. But the furious water hit as he reached the back of the truck. All he could do was grab it. Then he clambered in.

As the water kept coming, Ehrensberger floorboarded the old army vehicle. Surprisingly, it still functioned. The two firemen headed for the bridge — still hoping to sound

37

the alarm.

But it was too late. The raging water pushed the truck sideways and threatened to topple it. Since it had a canvas top, both men realized they could not stay in it.

Elsberry said, "We've got to get out of this," and threw himself out the door. Ehrensberger agreed, but since there was really nowhere safe to go, he was hesitant. As it turned out, he was hesitant to die. When he finally stepped out his hip boots apparently filled with water at once. The current immediately pulled his feet from under him and as he swept by Elsberry their outstretched hands missed by eight inches. Bill Ehrensberger was never seen alive again.

Meanwhile, Elsberry had grabbed a tree. But then the tree gave way and the current sucked him under. He says, "I took a big breath and I had one thought—get those boots off."

The fair-haired Nebraskan continues, "I felt power. I know the feeling—I've run heavy equipment, I've ridden bucking horses. But this was **power**. I got one hip boot off. I was on the ground underwater when something—I think it was the jeep—pinned my right foot. I thought, 'Is this how I die?' Then the object moved. I yanked and I came loose.

"I got my second boot off, even though I felt I was in a rolling barrel. I felt held under, and I thought, 'This can't be it, I'm too young to die.' But I felt peace.

"The clear thought came, 'You do what you can. If you get out, okay. If not, okay.'

"Then there was super peace—sensed, if not thought.

"At that point I ran into something on the bottom. It split my lip and loosened my teeth and as I came up, I was caught again. I was standing up, but still underwater, and glued to a rock.

"Then I came loose again, so I started swimming. I was short of air but I knew I was close to the surface. When I broke through I got mud—and a little air.

"I swam again, until I popped to the surface, gasping

38

and spitting. I knew that the creek had a bend and I felt I could get out there, so I headed for the bank. I snagged a vine but it started to give. To my right there was debris, so I scrambled up about three feet over the rubble, and found myself on level ground, just above the water but about sixteen feet above the normal creek level.

"I crawled three or four feet. I was tired and choking. But when the water started to rise again, I got up and ran up the road about twenty steps more.

"Then there was a terrific noise. The big gas main into the college had burst and the released pressure sent a geyser spouting thirty-five feet into the air." Somehow, in the drizzly darkness, Elsberry could see.

"I looked back across the creek and watched the Veer home collapse outward. The roof fell in. I had moved through all that water and now, in a few seconds, I was on the far bank.

"Then I remembered that Bill Ehrensberger had said his wife and children were all sleeping. I ran for their frame home, hoping to warn them, but halfway there I saw it begin to float away.

"The agony was awful. I knew a man's wife and children were floating away. Then the trailers started to go. They looked like tractor trailers pulling out on the freeway — they really took off and started to fly. They were okay unless they hit each other.

"There were many screams, mostly from children. I stood on the bank and watched people die, but I couldn't do a thing."

Trailerville had become a lake.

In retrospect, Elsberry says he had two big problems. Living with what his ears had heard. And understanding why. But time has changed his perspective.

Though Elsberry had been a minister before taking up his maintenance role at the college, he now says that the experiences during the flood have not damaged his Christian faith, but made it stronger.

One more thing. Elsberry did not live in Trailerville. His

39

family was safe on higher ground. He didn't have to be in
the path of this hellish water—but he was.

Chapter 5

I Didn't Want Daddy to Die

When Thad Fledderjohann, 3, got up the morning of November 6, he wanted to know why his mommie had been crying.

She said, "I was crying because Daddy was killed last night. But he's okay, he's with Jesus. I'm only crying because I miss him."

When her son Brian, 4, heard about the disaster he got angry. "I didn't want the dam to break," he said. "I didn't want Daddy to die."

Then they all cried together.

Brian's daddy, David Fledderjohann, was singled out as the hero who died in the Toccoa Falls tragedy. And he was that.

His last words to his men were, "Try to get back across the creek and sound the alarm."

Fledderjohann seemed to be drawn to emergencies. He inclined toward helping others and the six-foot young man from Ohio might indeed be embarrassed were he to know that the media paid so much attention to his heroism.

In the November 16 Memorial Service Dr. Kenn Opperman, college president, said it succinctly: "David Fledderjohann did not have to go into the flood area. He was there helping people because he felt he ought to go." And Opperman continued, "We believe the young people from Toccoa Falls will always be that way. They will be willing to go with the gospel of Jesus Christ—even if they don't come back."

Margaret Fledderjohann, David's widow, is slight, fair, blue-eyed and radiant. She had been an MK—a missionary's kid—in Nicaragua, the Honduras, and Mexico. Her Spanish accent is still authentic when she cares to use it. She met David at Toccoa Falls College and they were married at the end of David's junior year. Before graduation Dave joined the staff of the college print shop where he became the manager. He also liked paramedical and civil defense courses.

Dave was the first of his family to come south to Toccoa, but others have come since. The Greater Europe Mission once called him to go to Europe as a missionary. On another occasion Taylor University asked him to develop their print shop operation in Indiana, but because he didn't have God's peace about the matter he didn't go. Somehow he didn't feel his work was done. Nor was it—till the early morning hours of November 6, 1977.

Margaret says God has given her joy and peace after her loss. With dry eyes and deep earnestness she says it and it rings true.

The key to her present serenity probably goes back to an earlier time in her relationship with David. She remembers fearing that she would get a phone call informing her that David was dead. Believing it her Christian duty to give her husband completely to God, she did exactly that. There was deep peace then. And in her present bereavement, the peace remains.

It was raining hard the night of the tragedy. David had looked in the door about 9:00 p.m. He said he was watching the Sproull home and the bridge. Margaret put the children to bed and sat down to write two letters—

42

one to her parents, another to David's parents. She told them that it had been raining for a few days and that David was out on duty. When David came in at 11:30 they spent a last hour together, but at 12:30 he was called out again. The guard thought perhaps the bridge to Trailerville was going to wash out.

About 3:00 a.m. Margaret heard the telephone ring. The call was from a friend in Toccoa. "Are you all right?" she asked Margaret. Mrs. Fledderjohann was puzzled at the question, but said yes.

While she was still on the phone she saw the fire truck wheel in the driveway and heard a knock on the door. It was Dee Pinney, who came in with a shaken, hurt look on his face. He told Margaret that the dam had broken and that David was missing.

Margaret asked, "Was David on foot?" When she learned he was, she had little hope that he was still alive.

Dee had brought a couple of girls with him, who were waiting outside in the fire truck. He asked Margaret if she would like them to stay with her.

She agreed and Dee went to get them. Dee then left, but the two girls remained with Margaret the rest of the night. The girls told her they had heard the explosions from the fuses and also the rush of the water. While they waited, there was prayer, some tears, and the reading of the Bible.

At about 4:00 a.m. Doyle Fledderjohann, Dave's brother, came down to use the phone, since the Fledder- johann home was in one of the areas of the campus where the phones still functioned. He walked in so naturally that it was clear he didn't know that David was missing. But an hour and a half later he identified David's body at the hospital.

When informed, Margaret's first reaction was, "This is not real. David always comes back from helping people." Then Scriptures came flooding to her mind. "Heaven and earth shall pass away, but my words shall not pass away." "Absent from the body...present with the Lord."

Margaret Fledderjohann, still reeling from shock and sorrow, got down on her knees and impulsively said, "Thank you, Lord, that the dam broke." She was acting on a biblical principle—"In every thing give thanks."

And she prayed again, "Lord, teach me how to mourn. I can't sit down and analyze my feelings. Show me when to cry, and what to cry about."

Those tears continue to this day—but there's something else too.

"I knew David was in the presence of the Lord," she says. "Nothing has made it clearer to me that all we have been taught and have believed all these years is a reality.

"Joy and peace were the two most outstanding things that came to me. When I asked the Lord for strength, I got joy. I thought it was fantastic. The joy of the Lord really is our strength.

"Saint Paul wrote to Christians who were being burned and severed in pieces and persecuted in many other ways. And he said, 'Rejoice in the Lord always.' I've noticed the Bible is full of peace and joy passages—all addressed to suffering believers.

"The knowledge that God loves us and that anything that comes into our lives is through love is very precious. He does it in love."

The funeral in Toccoa's First Alliance Church was a celebration of victory. Ten victims of the flood, including David, were remembered.

Upon arrival at the cemetery, Margaret Fledderjohann asked church organist Nona Carlson if she thought it would be all right for them to sing at the service.

Mrs. Carlson decided to check with the officiating minister, Rev. Merle Graven. "Certainly," he said.

And so at the cemetery, upon completion of the prayer by Dr. Paul Alford, Mrs. Carlson began to sing softly.

It will be worth it all, when we see Jesus.
Life's trials will seem so small, when we see Him....

44

Some of the bereaved began to join in the singing.

A reporter/photographer from Athens, Georgia, couldn't understand why there was such joy. When asked if he would like someone to help him find what these incredible people had, he said yes. So in a praying, weeping huddle a newsman at a funeral called upon the Lord.

Still later a salesman who used to call at the print shop came by to see Margaret. He just wanted to tell her that through David's life and passing, he too had found the Lord.

Dee Pinney is a lanky Pennsylvanian, 26 years old, who worked with David Fledderjohann in the college print shop. He was also a member of the Volunteer Fire Department. In that capacity he participated in the creek-watch.

He also went in for coffee at the Ginthers'. Because he is interested in electronics he stayed a bit longer than the other firemen to tinker with the ham radio.

About 1:20 a.m. he left the Ginther home to cross to the other side of the creek. But as he drove across the bridge the campus power went out, so he set out to look for an electrician and a fire truck.

Unknowingly, he drove along the creek road—straight toward the descending flood. At that time the dam had probably already given way and the destruction was about to begin. For Pinney it appeared first as about six inches of water lapping down the road and against Forrest Hall. It was moving rapidly.

Instantly he thought, "The dam broke!"

He rammed his car into reverse. By then the water was to the floorboards. Soon the backup lights in the bumper were covered and he struck a tree. The water was knee-deep when he clambered out of his car, and almost immediately it was to his waist.

Carefully, very carefully, because he couldn't swim, Dee Pinney moved up the hill toward Stewart Hall, another large, frame building on campus.

About the time he reached Stewart Hall, he heard electrical wires falling into the water and then an explosion. He forced himself to work his way toward the center of the campus. The wall of water was storming by in all its raging power.

He headed for the fire department and as he crossed the campus a student asked if he should ring the iron bell. Pinney consented, since there was no electricity to sound any other alarm. Then, in a state of near-shock, he moved to the firehall, took out the truck, and sounded the siren as he headed for the hospital.

He tried to call his sister, Eloise, in Trailerville and thought he heard someone pick up the phone, but then it went dead. A nurse at the hospital called the ambulance and Pinney called 911, the emergency number, and told the operator to summon the Fire Department in Toccoa and also the Civil Defense.

Apparently, Eloise, Dee's twin sister, had called the telephone operator at about 1:30 a.m. She thought she had heard an explosion and imagined that the print shop was on fire.

When it became clear that a flood was involved, Eloise apparently left the trailer and headed her car for Upper Trailerville and safety. But the car must have stalled. In any case, Eloise did not escape.

She was a quiet girl whose off-campus job was at Sherino's, a local Italian restaurant. Many customers and fellow workers came to her funeral.

Pinney's first thought at the hospital was that he was the only fireman who survived. But when Eldon Elsberry came in crying, he knew he wasn't alone.

Dr. Opperman had been on the phone from Florida by that time so Pinney went to the Opperman home and placed a call. When Opperman came on the line Pinney told him what he could. "I'm sure some people are gone,"

46

he said.

Pinney proceeded to the Fledderjohann home to tell Margaret that Dave was missing. He couldn't bring himself to tell her that Dave was dead, though he felt sure that was the case.

He did not want to go back into the flood area. A lot of other help was on the scene by then and there seemed to be no need. He had been calling a friend in Toccoa, Mrs. Nona Carlson, who had herself been bereaved a couple of years earlier and was close to Dee and Eloise. Pinney ended up at Mrs. Carlson's home and it was from her that he heard the news that Eloise, his twin, was gone.

It was like losing a part of himself.

> And so, dear brothers, I plead with you to give your bodies to God. Let them be a living sacrifice, holy — the kind he can accept. When you think of what he has done for you, is this too much to ask? Don't copy the behavior and customs of this world, but be a new and different person with a fresh newness in all you do and think. Then you will learn from your own experience how his ways will really satisfy you.
>
> Romans 12:1-2, The Living Bible

II.
Residence Row

Chapter 6

Kitties in Heaven?

There were some little folks (five of them to be exact) at the Pepsnys', the first house on Residence Row. Three little visitors and two who lived there all the time. Kitties and bicycles were very important to them, but the Savior was too. What they didn't know was that the gates of heaven were already ajar.

Once past Morrison and Ralls, the wall of water, burdened with bobbing autos, loaded with debris, and probably moving at one hundred miles an hour, hurled itself through the narrow valley like a freight train out of control.

In the green and scenic valley, not much more than a hundred feet wide, sat Residence Row. Nestled beside the cheerful brook, it was the choice place to live on the campus at Toccoa Falls.

The houses there were frame and brick. One was gray. Another coral. Another rose. There were fences and cars. Clipped lawns and tended flowers. Boats and trailers. A mobile home with a frame addition was located there too. There were carports and shrubbery. Behind one house was an empty rabbit pen. A toolshed stood behind another. The street was improved.

It was this area of the campus that was to bear the full fury of the water. The first wall of water came through carrying boats and fences and trees before it. The water raced on to the bridge over Georgia's Route 17 where it was momentarily stopped, and though some of the water continued on toward Lake Hartwell, a backwash started back up toward Trailerville and Residence Row.

Observers on the scene believe that when the main wall of water, at least thirty feet high, lunged through the narrow valley it slammed into the smaller wave on its way back. The furious collision of those two waves hurled debris about one hundred feet above the normal level of the creek.

It was toward this sleeping campus suburb the howling water rushed that dark morning of November 6.

Edward and Carol Pepsny were new at Toccoa Falls College and their home was the first in Residence Row to face the onslaught. Ed was a graduate of Asbury Seminary in Wilmore, Kentucky and originated from Nashville, Tennessee. He had taken his undergraduate training at Case Western Reserve University in Cleveland, Ohio.

While he studied astronomy there with a college friend, the topic sometimes turned to religion. Christianity, Ed's friend decided, ought to be worth everything or nothing at all.

He soon decided it was worth everything and promptly sought to persuade Ed. One day he handed Ed a gospel tract.

Pepsny stuffed the tract into his pocket and proceeded to a Roman Catholic church to attend mass. Arriving a while before the service began, he drew out the tract, read it, and immediately made the same full commitment to Christ that his friend had made.

It was a major step on the path that led him to Toccoa Falls.

He had begun to teach at Toccoa Falls in January, 1977,

51

and had only moved onto the campus in the summer. Ed and Carol had two children — Paul, 5, and Bonnie, 2. Theirs was a quiet home and it was easily apparent that they were a loving, close-knit family.

Ed and Carol Pepsny were very private people, but yet hospitable. They had invited Debbie Jones, 19, to stay in their home and had she not been at work she too would have been lost in the flood.

Debbie and Carol had often talked about death and dying. About heaven, too.

Carol had once had a nightmare about her small boy drowning, and sometimes she seemed apprehensive about the creek. Debbie taught Carol and both the children to swim the previous summer. But it didn't, it couldn't, help, the morning of November 6.

Edward Pepsny still had not completed a full year of teaching Greek on the campus when the disaster struck. The night before the flood Carol Pepsny had felt a burden to pray instead of sleep. The Pepsnys did not know the reason for the desire to pray, they simply wanted to do so. They had also decided to fast on Saturday as a more earnest form of spiritual desire. Apparently, they were fasting when the waters descended upon them.

Ed and Carol were friendly with the Jerry Sproull family and when the waters began to threaten the Sproull home, Dr. Sproull took his three girls to the Pepsny residence. It was on higher ground and, he thought, not likely to be flooded at all.

The Sproull girls, Melissa, Jocelyn, and Joanna, cried a bit. Somehow they didn't want to spend the night away from their parents. But Jerry Sproull had been firm. He wanted to protect his precious children.

What he could not know is that he placed them squarely in the path of the raging waters. At 1:27 a.m., when the roaring flood descended, Ed Pepsny's little family and another man's dear children were all carried away. No one could survive the center of the rushing flood. And no one did.

Seven fatalities in one awful stroke. But surely angels attended the homegoing of a man of God, his wife, and five tiny believers.

Though hardly a brick was left, the little one who wanted to know if there were bicycles in heaven found out, and the tiny girl who wanted to know if there were kitties in heaven found out, too.

Suffer little children to come unto me and forbid them not: for of such is the kingdom of God.

Luke 18:16

Chapter 7

Something Wonderful

The rage of the waters was not assuaged. The next residence in the doomed row lay right in the path of the fury and was also angrily swallowed. It belonged to Paul and Mary Williams.

How "Uncle Paul" and "Aunt Mary" met death we cannot know, but we can surmise that they met it as they did life, with serenity and equanimity.

Uncle Paul was a Texan. He met Aunt Mary at Toccoa and though she briefly served as a missionary in Cuba she came back promptly to marry Uncle Paul. Just as promptly he took her back to Texas and stayed there twenty years.

But the pull of Toccoa Falls and the hearty charisma of R.A. Forrest proved irresistible. They came back. Uncle Paul was a rancher, a cattleman and a dairyman. But even in the days when a full-fledged farm operated at Toccoa Falls, he was more.

Uncle Paul was wise. Philosophical. Will Rogers-like. And dear to students, staff and faculty alike.

Aunt Mary was the kind who brought cookies to the guard shack for cold and lonely boys. The day before the flood she had given a married student, Jim Weiss, a ball of popcorn.

When there was a vacancy on the Board of Trustees at Toccoa Falls College, Uncle Paul was nominated.

Aunt Mary was the Director of Christian Service for women for more than twenty years. At the time she died she had in her purse $100 in Christian Service funds for the purchase of materials. When her purse was found after the flood the money was intact.

Aunt Mary not only directed Christian Service activities, she participated in them. At the hospital, in the schools, and in town, the doors were open because she cared for people.

Dr. Opperman affirms that sometimes at the end of the month when there was no way to meet the payroll for the college, Uncle Paul would walk in and lay a check on his desk. His resources were on the line for God.

When Uncle Paul was alive he always had appropriate stories, but he has a completely new story now. The grown children of Uncle Paul and Aunt Mary call them blessed. Their lengthy lives are over and their bodies rest on the hillside near the graves of Dr. and Mrs. R.A. Forrest.

The college that affectionately called them Uncle and Aunt honored them in death. They will be among the last to be buried on the tiny hillside.

Just the Friday before the flood, Aunt Mary saw the Rev. Nathan Penland, Executive Director, Alumni Affairs, in the campus post office. He may have looked downcast, because Aunt Mary said, "Don't get discouraged, Nate. I know we all do at times. Paul and I really believe God is going to meet the school real soon in a very unusual way. We don't know exactly what He's going to do, but it's going to be wonderful."

And wonderful it was! For godly people with heaven in view.

...no good thing will be withhold from them that walk uprightly.

Psalm 84:11

55

Chapter 8

I'm Ready to Die Tomorrow

A dozen dead? Yes, certainly, but the flood sought twenty-seven more. Dr. Jerry Sproull's home was next in the row, with no hope at all of withstanding the onslaught of gathering debris, the awful power of a wall of water thirty feet high.

When the water turned red, David Fledderjohann had gone to the Sproulls' to warn them. Ever the fireman at heart, as Ehrensberger had tried to get away in the jeep, and as the water surged upward, Fledderjohann urged the Sproulls to clamber onto the furniture in the house to keep away from the raging flood. But there was no escape and at the last Jerry prayed desperately, "O God—it's all in Your hands now." Little did he know that Melissa, Jocelyn, and Joanna, his little girls, were already in God's hands.

Pat Sproull had married Jerry in 1965. Their marriage lasted just a little over twelve years, but it was the kind of union that allowed Jerry to lecture regularly about family relationships. He had no fear that his audiences would be disappointed by his own marriage.

As the waters rose irresistibly in the Sproull home, Pat

placed her hand on the window and found it inexplicably open. The air conditioner had been sucked outward.

Then the waters pulled at Mrs. Sproull too. She is a slender woman but she remembers turning her head to let herself slip through the open window.

When she surfaced she put her hand down. Shingles! She thought, "This is the proverbial riding the roof!" But as she rode it she reasoned, "God must have something left for me. I've made it this far."

Then the roof collided with a demolished trailer. Again she was hurled into the water—but this time she grabbed a board, then a limb. Bill Anderson, another of the victims, finally pulled her out and she soon found herself huddled in the woods with a small group of survivors. They were slimy and in some cases naked, but there was singing, prayer, Scripture and, incredibly, some joshing and humor. Mrs. Sproull, though battered and injured, helped care for the children.

Pat Sproull clung to the Twenty-Third Psalm: "Yea, though I walk through the valley of the shadow of death, I will fear no evil: for thou art with me; thy rod and thy staff they comfort me. Thou preparest a table before me in the presence of mine enemies: thou anointest my head with oil; my cup runneth over. Surely goodness and mercy shall follow me all the days of my life: and I will dwell in the house of the Lord for ever."

Later, in the hospital, when the news came to Pat that all her children were gone, one nurse said, "Pat, how can you be so strong? You've lost all your babies tonight." Through her tears and brokenness she sustained her colleagues. (She is a supervisory nurse at Stephens County Hospital.) And she gave her friends an answer—"We gave our girls to God long ago, even before they were born. He's just taken them now."

Pat relates that Jerry used to talk about heaven. He looked forward to it. His inquisitive legal mind craved some of the details the Bible leaves unrecorded. Perhaps, she thinks, Jerry didn't think he would grow old. She

57

feels that had he known what was going to happen at Toccoa Falls, had he thought his death could have brought revival to the school or led someone to faith in Christ, he would have been willing to die. Jerry jogged and prayed every morning, and as he jogged he prayed for Toccoa Falls College.

Pat Sproull is struggling in her sorrow. She says, "I've lost everything. I don't know whether I'll ever be happy again. But there's joy. God gives me joy. There is a difference.

"Some books show only the bright side. And readers think, 'Those people can't be real!' But I loved my family. I'm crying and I'm crying still."

When America's First Lady, Rosalynn Carter, came to the campus Sunday afternoon, she visited Mrs. Sproull in the Stephens County Hospital. There, in Mrs. Sproull's room, the two women quoted the Twenty-Third Psalm together.

Some are weeping at Toccoa Falls and some are rejoicing. Perhaps that is as it should be. Pat Sproull still weeps.

But through her tears, or perhaps because of them, she has been able to reach out to others in a new way.

In her duties at Stephens County Hospital, Pat Sproull deals with life and death every day. Before the flood she always had felt unable to bring comfort to those who had lost loved ones. She used to busy herself with details to avoid dealing with bereaved people. Now, she assigns the details to others and seeks to minister comfort to the families.

Shortly after the flood she encountered a woman who had lost her husband. The woman was weeping and taking her loss very hard.

"I know how you feel," said Mrs. Sproull softly.

The woman looked at her as if to say, "How could you?"

Then Pat told her about November 6. The two women embraced and wept together.

"For the first time I felt I could help the bereaved," says

Mrs. Sproull. "For the first time I could feel for them."

Dr. Sproull may have sensed that the end was near. The day before the tragedy he was into one of his frequent theological discussions with a student—the same kind of discussions I had with Jerry when I met him at Toccoa Falls.

In the course of the conversation with the student he said, "I'm ready to die tomorrow."

And he did.

But before he died he had written an article and suggested that it be published anonymously. It was called "So That The Poor Mourn."

It was published in the Memorial Edition of **TFC Today**—but not anonymously, as Jerry hoped. The thrust of the article is that beside a man's open grave there should be a company of grieving poor who were helped in his lifetime. Jerry urged worthwhile living so that it might finally be said, "The poor were sorry when he died." The article's final sentence was, "Dear God, may the poor be sorry when I die."

Dr. Gerald McGraw, friend and colleague on the Toccoa faculty, could not help but respond, "We are sorry, Jerry. We are richer because you lived, and our loss is immense."

Blessed are they that mourn: for they shall be comforted.

Matthew 5:4

Chapter 9

La Suerte de Algunos

Betty Jean Woerner was secretary to Dr. Kenn Opperman and when I was a guest at Toccoa seven weeks before the flood, she was helpful to me as well. Since she had been a missionary to Chile along with her husband Ray, she knew Spanish and we tossed around some phrases. I distinctly remember her saying, "La suerte de algunos." It means, "The luck of some people."

What happened the dark morning of November 6 had nothing to do with fate or chance, Betty Jean would certainly affirm.

The Woerners had been missionaries with the Christian and Missionary Alliance in Chile for more than twenty years. On an extended furlough in Georgia, Ray was working in a nearby plant on the night shift. Their four children, all young people, were at home with their mother, in Residence Row.

At about 1:15 a.m. Betty Jean still had not gone to bed. She had a phobia about that water. Apparently there was no premonition, she just respected that creek, and seemed to be restless.

She went to her son David's room and told him to move the car away from the rising waters. He groused, half awake, but he pulled on his jeans and went outside. He also shoved a flashlight in his hip pocket.

At that time the water was one foot below the Woerner

house. After moving the car, David Woerner headed back to his home, intending to enter through the rear door.

Just then there was a sound like thunder, as the big fuses went out. Denise Woerner was wakened at 1:26 a.m. on her digital clock. She grabbed her clothes and was outside in about a minute. She saw the rising water and sensed immediately that the dam had burst.

As she ran for safety she screamed, "The dam's broken! The dam's broken!" David, in panic, started to run with her—until he realized that his sister Debbie and his mother were not with them. He dashed back to the house to get them. Betty Jean was speechless. Debbie grabbed one arm and David grabbed the other. They ran down the road, futilely as it turned out, because they stayed in the center of the flood path. David still didn't believe the dam had broken.

Denise had sensed it, though. As she ran through Trailerville she kept screaming and, according to one report, thumping trailers hard with her open palm. A number of people who were thus awakened were saved.

Somehow, David, Debbie and their mother made it to the print shop, a large, solid building, but squarely in the path of the flood. Never once did David think of the safety of the mountain off to his right. Daniel, David's brother, had joined them by this time, but the force of the still-rising waters soon pegged all four of them against the shop wall.

Daniel, a soccer player, sprang free and began to jump from one car roof to another. Like floating checkers, they led him to safety. He shouted for the others to follow and his mother broke her silence and cried out, "Daniel, be careful, come back." But Daniel didn't come back. And he escaped.

When the print shop began to crumble behind Betty Jean, David and Debbie, they pushed a car out into the current and moved out with it. Betty Jean clutched the rear view mirror. David held Debbie.

"Lord, I don't want to die," Debbie cried.

61

"Don't worry, Debbie," her mother comforted her. "It'll be all right."

Then the big wave hit, carrying everything before it. Debbie screamed and they were gone.

David was plunged under the water twice, then he grabbed a timber which carried him, at about forty miles an hour, straight toward the bank and an ugly pile of debris. But suddenly the current changed. He was not slammed into the cutting edge of the debris. Instead, he scampered over it to safety.

When Ray Woerner returned from work in a local factory shortly after 3 a.m., he found he could not cross the bridge on Route 17. There was water and debris everywhere. And a chilling sight that telegraphed disaster—his own trunk, which he had used on the mission field, went floating by. When Dave met him, he told his dad, "I don't know where Mom and Debbie are." But in his heart he knew they had died.

Grief for the Woerners has been hard, though it has not always been visible. David states, "I've shed tears twice—but a thousand times a day I've felt the emptiness."

From this deep hurt, however, has come an understanding. David believes Christians must weep with those who weep, rejoice with those who rejoice, and have the wisdom and discernment to know the difference.

He sums up his feelings this way, "There's a settled understanding, but grief beyond measure." And Betty Jean and Debbie? For them it was terror at Toccoa one moment, present with the Lord the next. That's what Saint Paul makes clear.

There are no shocks or bad luck for believers. But surprises, yes. Betty Jean and daughter Debbie, good and godly women, were surprised—by joy.

...absent from the body...present with the Lord.

II Corinthians 5:8

Chapter 10

The Baby Never Moved

Jaimée Veer was a breast-fed baby and according to Sue, her mother, she was a very loving child. She cuddled a lot, always searching for her mother's breast. And she was terrified of water.

But the night of the flood Jaimée did not move. She never searched. She never cried. Incredibly, in all of the turmoil, even in the water, she was quiet.

"I believe," Sue affirms, "that God may have taken my baby before the flood came.

"My greatest memory is that of the soft, warm body against me."

Earlier, the phone had rung at the Veer home. Sue's husband, Doug, was a supervisor of maintenance at the college and the call would have taken him away from his family, away from the row of residences lying silently in the flood path. But for some strange reason, no one heard the call and it went unanswered.

About 1:30 a.m. there uas a different sound, which they did hear. It was somewhat tinny and hollow. Others have described it as two oceans rushing to reach the same place. Doug jumped up and looked out the window.

What he saw looked like a train—but there wasn't supposed to be a train in the valley. Apparently, it was the big wave coming through.

The baby Jaimée was sleeping with her parents. Doug pushed both Sue and the baby up the ladder leading to the attic. By then he was up to his chest in water. Anita, their 14-year-old, was screaming, but the baby did not waken. As the water kept rising, Doug cried out, "My God, I can't get to them all!" He still had not reached his two sons.

Sue heard Doug go underwater. She says, "I felt alone. I didn't want to be alone or to die alone." Their frame house with a brick front began to shake, so halfway into the attic she doubled over the baby and covered her.

Suddenly a tree rammed through the trembling house, knocking her off the step. But still Sue Veer clung to her little girl. Underwater, Sue could hear the noise of the flood. Still, the baby did not move.

When Mrs. Veer was hit by either a piece of concrete or a tree, she and the baby were flung apart.

Sue tried to fight clear, but she lay under debris. Finally, she took a deep breath. It should have been water but, unexpectedly, it wasn't. She just blacked out. Then she came up, churning but not coughing. And without water or mud in her lungs.

Suddenly, her face was caught and began to be crushed. She could hear the facial bones cracking. "It was sickening," she says. "I wanted to vomit."

She still couldn't believe what was happening to her. She found herself thinking about the Lord and His suffering on the cross. Then suddenly she was dragged down again.

Next, Sue remembers being on her back underwater. She saw lights above her and she assumed she was dead. As she floated upward she expected to meet Jesus at any moment.

But when she broke surface she encountered debris and a layer of mist and gas fumes several inches thick clinging

64

to the waters. She also smelled the reeking odor of escaping gas mixed with mud. She knew she wasn't in heaven yet.

The water was still moving fast. It seemed like fifty or sixty miles per hour. Somehow, she thought she was going over the falls.

Whenever an object went by her she shouted, "Get to the shore!" One object answered her. It was her daughter Anita, who immediately dove off the flotsam and came to her mother. They were both naked but alive.

"Momma, we've got to pray," Anita said.

They clambered out of the water and debris in the area of the old farm, at the far edge of the floodplain below Trailerville.

"Momma, tell me I'm asleep," said Anita.

But Sue couldn't.

Sue's jaw and facial bones were broken and her attractive, cameo-like face was badly swollen. Later, when she was being carried on a stretcher to the hospital, Doug shone a flashlight into her face. He didn't even recognize his own wife, and he said to himself, "That woman can't live."

The medical examiner suspected a broken left leg, spinal injuries, a crushed skull, a broken neck, three broken ribs, and ten broken facial bones.

But as it turned out, after prayer, Sue had only the facial injuries. The other injuries were gone.

When Doug went underwater, he and Sue had become separated. The tree had come pounding through, knocking out the end of the house with frightening force. Doug, Jr., 18, had cried out, "My God, Dad, help me!" Kirk, 10, had somehow gotten tangled up in a big flag that had fallen off the wall. Later he shot free.

Doug, Sr. says now, "All of us were underwater. We all ran out of breath. We all took a huge gasp. We all should have taken in water and mud but God gave us air, and though we all blacked out, we came out without choking."

65

Doug hit something solid underwater. It turned out to be a forklift. Somehow, the top part of it was still above water.

Previously, Doug had been asked what he would say if God took all his family from him in an accident. He had surmised that he really might blame God to His face in such a situation.

But Doug had misjudged himself. When he broke water at the top of the forklift he cried out, "Dear God, I love you."

For twenty-five minutes he clung there. As forty-five-gallon drums floated by, Doug was tempted to grab one of them. He's now glad he didn't, because he suddenly heard crying and talking.

It was Kirk, his 10-year-old son. "Oh Daddy," he said, "I'm so cold."

Kirk had felt the door handle of a truck come into his hand underwater. Instinctively, he opened it and climbed into the bubble of air. There he survived, close to the forklift his father had seized.

He had been praying, "Thank You, God, for the water. Thank You for the flood." And he had wanted to know if it was okay to pray that way. His father assured him, "That is what the Bible says — 'In every thing give thanks.'"

Kirk went on, "When I was finished praying, I intended to jump back into the water and go to be with Jesus and the rest of you." To him, it had just seemed to be the obvious thing for a believing little boy to do.

Kirk and Doug stayed together and the water began to recede. Fifteen minutes later they walked out. Doug, Jr., climbing onto a broken trailer, had escaped too.

Later, Sue had another prayer answered. She wanted very much to attend Jaimée's funeral at Toccoa's First Alliance Church. Though she was confined to the hospital, she saw even the committal service on television. The television clips seemed to be specially edited for her.

"We feel honored to have gone through it," Sue says.

"We feel God chose carefully those who were to suffer and still give glory unto Himself. Jaimée was ours. But she is gone. God gave us Jaimée long enough to teach us how to love one another."

Many waters cannot quench love, neither can the floods drown it....

Song of Solomon 8:7

Chapter 11

Anguish Ripped at Him

The night before the flood, the Ron Ginther family had enjoyed a special time in which the parents had read from **The Silver Chair,** an allegory by C.S. Lewis from **The Chronicles of Narnia**. It was a tale in which the king died and was raised again.

Ron and Mary Jo had questioned their four little girls carefully. And the children had understood the allegory's message. What they did not know was that before the night ended, all but Ron would perish in the flood — to await with certainty the coming resurrection.

Shortly after the firemen's coffee break at the Ginthers' home the water began to rise dramatically and alarm seized Ron Ginther. He ran to his trailer home.

"We've got to get out of here!" he told his wife. "I think the dam broke."

But by the time Ron and Mary Jo had aroused their children, the water was already to their knees. In a desperate attempt to escape, Ron kicked out a window, but his flashlight beam fell on gushing water. And inside the mobile home, the water continued to rise.

When the trailer began to float, Ron knew they had to

get out and a roof section floating by seemed to be the answer.

He placed one of his little girls out on the roof — but she was immediately washed away. Awful anguish ripped at Ron Ginther, but there was no time to mourn their terrible loss.

His wife urged him to go out on the roof himself so he could help the rest of them.

The children had been shouting and screaming, but Mary Jo had assured them, "Jesus is with us, He'll take care of us." Her serenity was remarkable; she was ordinarily terrified of water.

With his second girl, Rhonda, under his arm, Ron struggled through the window, only to be pinned between his house and the roof. He thought he was going to die right then and remembers praying, "Oh Lord, if this is what You want, I'm ready to meet You."

Suddenly, the buildings which were squeezing the life out of him eased apart.

He still had Rhonda under his arm, so they both clambered onto the roof. Then he turned toward his trailer.

But it was gone! Where once there had been love, warmth, a beautiful family — nothing!

Residence Row was gone. Obliterated.

Back of that row, pressed in against the mountain, Ron Ginther had tucked his trailer. The mobile home, with a frame addition, lay right in the path of the thundering flood — an onslaught loaded with boulders, trees, even bouncing pianos swept out of the music building upstream. Now it was no more.

So far, Ron's struggle to save his family had been a losing one, but he still had Rhonda. Then the roof on which they were precariously perched began to tip, and violent waves began to pound it harder and harder. Finally, a mighty wave crashed into it and his little girl was torn from him.

Ron, himself, though hammered by the water, managed to hang on.

69

He clung there till the fury passed. Later, he was able to walk out in knee-deep water. When Ron joined a group of students and told them his whole family had been drowned, they broke into spontaneous prayer.

Finally, he headed for the hospital. He knew within that he had lost his whole family, that Mary Jo, and his four small daughters, Brenda, Rhonda, Nancy and Tracy, were gone. Yet he still had peace. "The feeling," he says, "was the most real in my life."

Ron Ginther is still a student at Toccoa Falls College and is looking toward the ministry.

This soft-spoken man, who has lost so much, buried his little family not too far from Toccoa Falls. This was home, and they were happy here. And he says, "If we had it all to do over again, even if we knew the flood was going to come, we'd agree to come here again. This is where God wanted us—I know Mary Jo would feel this way."

Ron Ginther is not trying to be sensational. "We've only been Christians since 1975," he says. "This is how it really was."

For this God is our God for ever and ever: he will be our guide even unto death.

Psalm 48:14

III.
Tragedy in Trailerville

Chapter 12

There Is Still Pain

"The hardest thing to deal with is absence," says Jeff Metzger, 29.

"There is still pain. Cassie and Dirksen are not coming back."

There are little reminders that make it hard. It hurts when Jeff sees little children Dirksen's age. On a visit to Pennsylvania, when he saw a couple just brushing snow off a car, he felt it keenly. Cassie and he can't do things together anymore.

Cassie had always told him, "Jeff, you'll have to endure trial. Your ministry won't be easy."

Just past Residence Row, Toccoa Creek takes a bend south and to the right. The mobile homes on a small rise across the creek and the cluster of trailers in the heart of "Trailerville" itself all lay right in the path of the maelstrom.

Trailerville really had a more pretentious name, Falls Villa, but somehow that name had never stuck to the small community.

As the general residential area for married students at Toccoa Falls College, Trailerville had its own mayor and council. It was a haven of fellowship, a place where a kind of student subculture found its role in the larger college family.

There was a gravel road winding through and trailers were parked in orderly fashion, some newer, and some older. The creek, too, passed right through Trailerville. There also was a second level called Upper Trailerville which was untouched by the flood.

The students visited back and forth. They talked about cars and children. About power tools and trailer skirtings. About sports and term papers.

They talked about the Lord too. They prayed for each other, loved each other, and baby-sat for each other. They found it tough financially to get through school and also rear their families. "In Trailerville," one of the men recalls, "we were all broke together."

When the flood lurched around the bend in Toccoa Creek there was havoc everywhere and a thousand things happened at once.

Twenty-one were dead or dying — but the voracious flood demanded eighteen more, many of them children. All from Trailerville.

The events at the Jeff Metzgers' were similar to many other homes that fateful November night.

Jeff was working on his campus security detail. The lights were out, so he wasn't able to continue to study or read in the guard shack. Back at his trailer a candle burned.

The guards drove around the campus, checking and looking. With no lights to study by, some of the students were out parading around, having fun. The guards stopped and hassled them, also in fun.

Down at the bridge to Trailerville the water-watch was on. The water rose and fell. Some cars were moved and there was idle talk, but no apprehension. And no inkling of what lay ahead.

But at 1:30 a.m. the Metzgers were awakened. People were screaming, "The creek is flooding!"

73

"So what?" was Jeff's response. But he grabbed a flashlight and as he pulled on his trousers he prayed, "Take us through, Lord."

Suddenly he felt his trailer begin to move. Then he heard a loud bang outside. Metzger dashed into the bedroom, grabbed Dirksen and handed him to his wife Cassie. He picked up Debbie, his little girl, but when he tried to get the door open, he couldn't.

He hit it heavily with his shoulder and it sprang open. As they moved outside, the water was already thigh-deep and rising.

Cassie was making sounds of distress and Metzger couldn't understand why the water was so high.

There was a small creek between Metzger and the safety of the hill just adjacent. The family still had both children. Then Jeff fell into a submerged gully. He scrambled out on the trailer side, still as far away from the bank as ever—and moving dangerously close to the ugly current.

Then Jeff looked back. Nothing! He couldn't believe it. There was water all the way to the far side of the valley! The trailers were gone. He just couldn't believe what he saw.

A car zipped past. Its speed in the water seemed to be at least thirty-five miles per hour. Strangely, though, where the Metzgers were there was no current.

They saw another student, Bob Harner, and tried to walk in his direction. After about twenty feet the current got very swift. The Metzgers tried to go back but couldn't, and in that instant Jeff knew they were in deep trouble.

Then Metzger and his family lost their footing completely. Probably they had walked into the smaller creek that fed the one in flood.

Jeff, with Debbie in his arms, went under and he knew that Cassie and 2-year-old Dirksen had gone under too.

Metzger vividly remembers fighting and striving to keep hold of his little girl. They were slammed by debris, and Debbie, who was looking over his shoulder, said it was

74

their trailer.

At that point, Jeff felt a tremendous pressure on his chest. He thought he was going to die right then and hoped the pain would not be too severe.

But, the next thing he knew, he was shoved, pushed, and swept toward the bank. There he grabbed the bushes. Somehow he still had hold of Debbie.

Then something big and very heavy settled on his legs. He thought, "Oh no, I've come this far only to have my legs cut off!" But just as suddenly the pressure began to ease.

Later Metzger returned to that spot. He found a rock slab about ten feet long, with a big groove in it and fresh marks. He now thinks it may have been tottering back and forth under the water and that for a moment it had pinned him.

When the pressure had released, though he was badly bruised, he tossed Debbie to safety and climbed out himself. She also was severely bruised, apparently where Jeff's fingers and thumb held her in an iron grip.

It was quiet then, and he rejoiced that God had spared his life. But both father and daughter knew that the rest of the family had been lost.

Then he heard screaming from a grove of trees out in the current. Metzger charged back into the churning water from which he had just emerged, and swam out across the current to where 10-year-old Kendra Smith was perched. A sturdy, strong girl, she was unclothed and screeching like a jaybird.

Jeff grabbed a mattress that had jammed nearby. Then he crossed the current again, pushing the mattress and Kendra to safety.

Jeff recalls the time when Dirksen had been dedicated to God in a church ceremony. Cassie had agonized. She had cried and cried, but finally she had surrendered him to the Lord.

"God must have wanted Dirksen for some reason," Metzger continues. "I think God took Cassie because she couldn't have lived without him. And God never allows us to suffer more than we are able to bear. I think that is why He took Cassie too.

"Debbie and I can stand it. So what if we are left behind? We haven't asked why. We haven't cared why. God has His purposes."

> Be not thou therefore ashamed of the testimony of our Lord...but be thou partaker of the afflictions of the gospel according to the power of God; Who hath saved us, and called us with an holy calling, not according to our works, but according to his own purpose and grace....

> II Timothy 1:8–9

Chapter 13

I Want to Die in America

The man in the cockpit of the bomber was new and something had gone wrong. Frantically he pushed buttons—including the one that released the bombs.

The high explosives tumbled out on the tarmac in Thailand.

But they did not explode.

If they had, Bob Harner, 25, who was working nearby, would not be in Toccoa Falls College today.

But he's here and before the flood he had a Thai wife, Tiep, 22, and a baby boy, Robbie, 2.

When he met Tiep she was a slave in Thailand. The Thai didn't like him to say that, but since she had been bought and sold for money, in his eyes she was a slave. He took her to the Thai police and liberated her; five months later he married her. He had also learned to speak Thai.

Bob and Tiep didn't love each other before they were married, but they knew that love is learned, and Bob now believes that they achieved it.

When Tiep came to America she was especially impressed by the beautifully kept cemeteries. Apparently, in Thailand they are not as well kept, because Tiep had said, "I want to die in America."

Since there were no lights the night of the flood, Bob and Tiep had gone to bed early. Tiep's wish was about to come true.

Just before the flood descended upon them in all its force Bob and Tiep had heard screams — "Get out! Get out!" There were also car horns blowing, and Bob immediately figured, "Flash flood!" But by the time he had shaken Tiep four or five times, the water was already moving the trailer off its blocks. When the water hit the trailer, both Robbie and Tiep screamed, but he managed to calm them.

Harner didn't want to stay in the trailer because he saw that other trailers were breaking up. He also noticed that Jeff Metzger, another college student, and his wife Cassie were carrying their children and walking toward the mountain. Apparently they had forgotten about the deep ditch right by the bank. The Metzgers did not respond to Harner's warning cry. Then, suddenly, they were swept away. Bob missed an outstretched hand by six inches. It might have been Cassie Metzger's last desperate grasp for life.

Outside their trailer, Bob and his family clung to the top of a car, but then a backwash of some kind swept them off — Robbie and Tiep one way and Bob another. He never saw them alive again.

For a few moments the violent water tossed Harner between the trailer and the car. Then, somehow, it swept him through a hole that had been punched in the mobile home.

The current sucked him down and kept him there. He was being dragged along the bottom. Three times he planted his feet and tried to leap to the surface. Once he felt nails puncture his feet like a knife slices into an orange. But it didn't hurt and as he kept trying to swim, passages of David's psalms raced through his mind. David had cried for deliverance long, long ago and Harner needed deliverance now.

Bob had been converted to Christ as a boy of twelve,

An aerial view.

Gate Cottage.

Flood marks on the trees.

Piano afloat.

The bodies washed to the bridge.

Ray Woerner being interviewed.

Press conference. Dr. Opperman (partially obscured by the cameraman) is flanked by General Billy Jones and Governor Busbee. Dr. James Grant is standing directly behind President Opperman.

The New Campus

New facilities.

The student center.

The enlarged Forrest Hall.

then had wandered from his faith in Thailand. But when Robbie had "died" at birth and then had been revived by cardiac resuscitation, Bob's heart had turned toward the Lord.

When his lungs were nearly bursting, something pulled Bob upward. He doesn't know what or who it was. When a floating, bobbing car came by, he seized it and it carried him to the old farm where he scrambled out of the water.

"I knew Robbie was gone and the thought came to me, 'Suffer little children to come unto me, and forbid them not.'"

Bob feels that he and his family came to Toccoa Falls College in the will of God. The remarkable manner in which they sold their house for twice the price they paid for it seemed a strong indication of God's direction.

Early that fateful morning, Robbie's body was found. But Bob kept asking for Tiep. Although someone told him she was okay, he felt uncertainty. Later he went to Stephens County Hospital to identify Robbie's body. When he pulled back the sheet, Tiep's body was there too.

Bob had been a member of one of the small clusters of survivors that had emerged on the south side of the floodplain. "The thing we did up there was praise God," he recalls. He especially remembers singing the chorus "God Is So Good," and sometimes he sings it now, with tears.

Harner is thoughtful and philosophical about his loss.

"Really, I didn't lose anything," he says. "My greatest responsibility as a husband was to see my family come to faith in Christ. My family knew Jesus. They are with the Lord.

"My concept of heaven is different now. Death used to be a place, now it is a door. I now see heaven as a place for worship, not a school in which God will answer questions.

"There was love in Trailerville. I remember Tiep looking

for something special to give to Cassie Metzger, our neighbor. The memories still hurt."

Despite the memories, Bob is continuing classes. He has moved into the men's dorm, where he has a roommate who is gradually losing his sight.

The roommates find strength in one another and Harner affirms, "We have a great relationship.

"Of course," he adds softly, "my roommate now is not quite as good as the one I had."

> *In my Father's house are many mansions: if it were not so, I would have told you. I go to prepare a place for you.*
>
> John 14:2

Chapter 14

Momma Starts to Pray

Mike and Ruth Ann Moore had bought a new bed Saturday, November 5. They had been to town and splurged. That night they stayed up till nearly midnight visiting their neighbors.

But just before 1:30 a.m., somehow sensing that something was wrong, the Moores and Ruth's sister, Donna Porter, a guest in their home, all got up. Baby Jeremiah was in the trailer's addition. The new bed had been used only an hour and a half.

When the Moores looked out they saw about six inches of water on the ground. They got Jeremiah and headed together into Mike's study.

Suddenly the front of the trailer was torn away and what was left was spun around. They could see a massive wall of water coming. It was ten feet over the top of the large bus sitting a few yards away, and it slammed their mobile home into a clump of trees. Then something smashed them from behind.

Mike says, "We were thrown into the water. I was never on top. It seemed that I was under all the time. I came up twice and I tried to swim. I was very scared and I don't think I was ready to die. I tried to stay alive because I didn't know if Ruth and Jeremiah had made it." The water carried Moore nearly a quarter of a mile, toward the bridge over Route 17.

Once out on the highway, Mike somehow knew that Ruth and Jeremiah had been lost. "Had I known they were both gone," he says, "I probably would have given up. They seemed to be first in my life, not the Lord.

"But there's been such a change since. Now I would be ready to die.

"God allowed the flood," says Mike. "The devil wanted it done, but now he wishes it had not happened. God has been getting praise and love and glory."

When Mike was thrown into the water, Ruth Ann, Jeremiah and Donna were somehow held back. Suddenly they too were swept away. They got to a log where Donna and Ruth sought to give mouth-to-mouth resuscitation to the baby. Donna started to take Jeremiah when Ruth said, "No, give him back to me, we're going home to be with Jesus."

Suddenly Ruth and Jeremiah were sucked under the water. Gone.

Mike continues, "Ruth had grown so much spiritually in the past year. She knew more. She was more ready. I don't think I could have reared Jeremiah. But now I know he's taken care of.

"It's hard. I want them back, but I have to go on.

"Ruth had prayed so hard that her family would find the Lord. After the tragedy, I went to visit them.

"I drove into the driveway and I just walked into their house. I found Momma sitting on a chair and I simply said, 'You know what you have to do.'

"'Show me how to pray,' she answered. Ruth's mother came to the Lord that day, and Momma's life has been different ever since. There have been others who have turned to the Lord too."

Mike Moore is broken and hurting. He's not one of the

radiant ones after his loss. But he's coping. He's weeping. And as he told me his story, I wanted to weep too.

> *...the Lord gave, and the Lord hath taken away; blessed be the name of the Lord.*
>
> Job 1:21

Chapter 15

Impaled by a Plank

When Mrs. Bea Rupp regained consciousness, she was a quarter of a mile downstream — pinned down by a roof, and with a two-by-four nailed into her thigh. The pain was excruciating, but when she tried to escape from under the roof, the plank kept her impaled.

Fortunately, the water there was shallow and quiet. As she lay there helplessly, she could hear the screams of adults and children and the whining of dogs.

Earlier that evening, Mrs. Rupp and her husband, Monroe, a retired minister, had gone to bed to a background of sounds far different. Heavy rain had been beating against their trailer roof for three days. But since their home was located on a bank above the creek, they were unconcerned.

"At 8 p.m.," says Mrs. Rupp, "the lights went out, so I found a small hurricane lamp and a larger oil lamp. We felt quite cozy, though the lamps did not provide enough light for us to read from our Bible, as we usually did. Instead, we quoted some of the chapters we had memorized. Then we sang together.

"My choice was to sing the chorus that ends with the

words, 'Sing Hosanna to the King of Kings.' Monroe chose to sing, 'Oh, How I Love Jesus.' After a time of prayer we went to bed.

"At 1:30 a.m. we were suddenly awakened. We jumped out of bed and rushed to the back door. The sight we saw was horrifying! Ralls Hall was afloat and speeding down the swollen creek, and behind the dorm was a wall of water thirty to forty feet high. Immediately, we knew the dam above the campus had broken.

"Monroe said, 'Quickly, close the window.' Then he took my hand and rushed toward the front door, but even as we ran down the hall, the water reached our ankles. From then on, everything happened very quickly. It was already too late to escape as the huge wave came upon us.

"The last thing that I heard my husband say was, 'Hang onto me.' As we clung to one another, I felt the tremendous force of water hit and break our mobile home to pieces. Then I lost consciousness."

After regaining consciousness, while still impaled by the plank, Mrs. Rupp began to pray audibly and to call for help. She kept it up for three hours. During that time she heard something upstream dislodge and she knew it would once again set the debris moving. "I felt certain I'd be crushed against the roof holding me down," she says. "But then I sensed a voice within me say, 'Move to the right.'

"I knew I had to obey that inner voice, so I squirmed my shoulders over as far as possible, still pinned down from the thigh. The debris came crashing down past my left side.'

Bea Rupp, struggling with pain and nearly immersed in stinking water, had no choice but to wait.

"Finally, after three hours in the cold waters, one of our students found me and sat with me until rescue workers came. In the darkness he had heard my voice and come to me.

"My husband went home to be with Jesus that early

110

morning. The parting has been extremely hard, but I'm reminded of the first verse of our wedding song:

"All the way my Savior leads me,
What have I to ask beside?
Can I doubt His tender mercy,
Who through life has been my guide?
Heavenly peace, divinest comfort,
Here by faith in Him to dwell!
For I know, what e'er befall me,
Jesus doeth all things well."

It has been a long struggle for Mrs. Rupp. Her wound was badly infected by the polluted water. But surgery four times has not dimmed Bea Rupp's enthusiasm to do God's will, nor her conviction that He does, indeed, do all things well. She told us over the phone, "The Lord is becoming more and more precious to me."

"We've been saved four years," say Thurman and Dixie Kemp, who come from Florida. They aren't telling about being rescued from the flood, but they are describing the religious conversion that led them to study at Toccoa Falls College.

"Ours was the first trailer to go, but the five houses were already gone," says Dixie, an attractive woman with dark brown eyes. She is referring to Residence Row.

The Kemp home was set on a little hill above the creek, but certainly not high enough to save them that fateful morning.

The evening before, all four of the Kemps had gotten into their king-sized bed. Chris, just 7, had read a story. Morgan, his brother, 9, had read one too. It had been a beautiful family time. The spirit had been unusual. Later, as Dixie lay in a hospital bed, still not knowing that Thurman and Morgan had survived, she thought, "God gave us a very special time."

Mrs. Ruby Kemp, Thurman's elderly mother, had been a guest in the home and when awakened about 1:30 a.m. she began to scream, "It's a tornado!" Her warning saved the lives of the Kemp family. A non-swimmer, she herself rode a piece of furniture to safety.

When Thurman was aroused, the water was already two feet high on the door. He managed to get to the back of the trailer as Morgan ran to him. Dixie tried to waken Chris, but he seemed to be leaden on his feet.

Suddenly, in the turbine-like thrashing of water, and in the reeling movements of their floating trailer, the floor collapsed. Dixie felt Chris being pulled away from her. Twice she pulled him back. But the third time he slipped out of her grasp. "I think now," she says, "that Chris was already gone. He felt so heavy.

"I must have been hit, I 'came to' underwater. But there was nothing beautiful about it like I had heard there was supposed to be when a person enters heaven.

"Come on, Lord," she prayed, "where's the beauty?"

Then she was struck on the head and temporarily lost sight in one eye. "Lord," she prayed, "drown me or get me out of this—but do it quick!"

Again on the surface, Mrs. Kemp looked to the left. Through the debris there was a little path. She climbed out.

"I think an angel pushed me out," she says now. "Then I thought about my family. How could I live without them? I thought about jumping back in."

But a phrase from the Scriptures slipped into her mind: "beauty for ashes." And now Mrs. Kemp says, "The Lord has given me just that."

Dixie Kemp was the first of her family in the hospital. She'd had the quickest trip.

Thurman and Morgan had been pulled in another direction. They were moving, Kemp feels, in excess of thirty-five miles per hour.

112

"We were swept up against a hill, where an undertow pulled us apart. Then I grabbed a tree, turned around, and I saw Morgan again. Instantly, I grabbed him. It was a miracle that I turned around when I did, and that he was right there.

"We clambered out on the hillside and then climbed higher. Morgan had a big hole in his leg, but it didn't bleed. He didn't get scared and he didn't cry for three and a half hours."

"By 5 a.m. we knew that Chris was gone," the Kemps state. At 9 a.m. his body was identified.

Despite their loss and the grief they felt, both Thurman and Dixie cooperated with the probing newsmen. One questioned Thurman about the earthen dam that had taken his son and so many of his friends.

He replied, "If God had wanted to hold that dam together He could have done it with a Band-Aid."

When another reporter asked him how he coped with the tragic situation he answered with the New Testament: "All things work together for good to them that love God, to them who are the called according to His purpose."

The day that Chris was buried, an aunt and an uncle, who had travelled to the funeral in separate cars, both declared that they too had placed their trust in Jesus Christ. The death of a little boy had drawn them into the fold.

Dixie concludes, "My little boy's funeral was full of grace and glory...I really feel that those lives that were taken were an offering to God, a sweet-smelling sacrifice to redeem the school and to reach out to the world.

"I think that God is trying to say, 'Hey, world, wake up, I love you.'"

We have a feeling that Dixie Kemp's conclusion is exactly right.

Unto you therefore which believe he is precious....

I Peter 2:7

113

Chapter 16

Get Out! Get Out!

"Get out! Get out! The dam has broken!" It was Denise Woerner running through Trailerville, sounding the alarm.

Bill Anderson, 33, a stocky New Englander, fumbled open his trailer window and heard the words clearly.

He thought there might be water—but he had no idea how much.

By the time he had pulled on his trousers, his wife had tumbled their five children out of bed and had herded them into the hallway.

Bill explained to his family, "There's a flood!" Then they started to pray, asking God to keep them from panic and to give them strength.

At that point, the trailer began to move. Almost simultaneously, a tremendous force slammed against it. Anderson now thinks it may have been the big wave.

The roof was torn off and inside there was turmoil—but somehow the Andersons got their children calmed down. And they encouraged them to trust in the Lord.

Because the hallway seemed to be the safest place, Mrs. Anderson and the two youngest children stayed there.

Bill stacked up furniture and climbed up to peer out through the slash in the roof.

A second crash came, ripping off the end of the trailer. This time, instead of panicking, the children cried, "Jesus! Jesus!"

Bill observed that they were nearing trees. When their trailer lodged against them he tossed his oldest daughter, 12-year-old Lisa, and his oldest son, 6-year-old Billy, into the branches of the trees.

Bill's wife, Karen, stayed in the hallway of the sinking trailer home. Bill heard her say to the children, "Come on, kids, get ready. We're going to meet Jesus." She was getting them prepared.

Clinging to the top of the trailer, Bill turned in time to see the wall collapse inward—right where his wife and two of his children stood. There was a swirl of water and they were gone. Anderson himself was thrown into the water. Desperately he clawed and scrambled over the debris toward the trees—and not once did he get wet above the shoulders.

But he knew his wife and two youngest children were with the Lord. "Thank you, Lord," he prayed, "they don't have to suffer anymore."

But uncertainty about his other daughter, Susan, persisted. She hadn't been under that wall. He didn't know where she was.

"I did have a sense of the presence of Jesus," he states. "It seemed as if the Lord had just reached down and taken my wife and children. There was peace that calmed my heart."

Clinging to a tree, with his feet on a wire, Bill Anderson saw flashlight beams. There were people in the woods. Dave Woerner helped the surviving Andersons to the shore and then he and Bill began to lead other survivors to their group. Because of the nakedness of some of them they said, "Huddle together. Keep warm. Put your arms around each other and lean against each other."

Then Anderson heard cries.

115

"Daddy! Daddy!"

It was Susan.

"Is that you, Susie?"

"Yes. Can you help me?"

"Where are you?"

"I'm on a piece of tin."

"Is it a big piece?"

"Yes."

"Is it moving?"

"No."

Anderson ordered his daughter to stay put. And later Susan was rescued from the other side of the swollen creek.

When Anderson's relatives came south from Vermont for the funeral service, Bill took one look at his sister. Somehow, she looked different. Soon, in a private moment, she told him, "I've found the Lord."

"I knew it," Anderson responded.

Later, when they went for a drive, Bill's oldest brother said, "I didn't come here for the funeral."

"What's the matter?" said Bill.

"I'm here to see you. I want what you've got." Bill Anderson led his brother to faith in Jesus Christ.

Karen and Bill Anderson had prayed for conversions in Bill's family for years. A Toccoa friend said of Karen, "She had redemption on her mind. She would gladly have died if she had thought it would bring others to the Lord."

Asked why there seemed to be serenity in the face of such grievous loss, Bill Anderson says, "The knowledge and the certainty in our hearts that these loved ones are with the Lord now, and that the Lord is present with us now, makes all the difference."

In our interviews with the survivors of the Toccoa Falls flood, we consistently observed this serenity. Not without tears and sorrow necessarily, but quiet joy.

And we feel obliged to agree with Anderson. This

knowledge does make a tremendous difference.

...I count all things but loss for the excellency of the knowledge of Christ Jesus my Lord: for whom I have suffered the loss of all things....

Philippians 3:8

Chapter 17

The Bus Didn't Stop

The old college bus was floating along in the flood tide as Jerry Nicholson and his wife, Pat, clung to a mattress on which they had placed one of their 7-year-old twin boys. The other twin clutched the mattress too. Jerry looked up and said, "Here comes the bus. I wonder if it'll stop!"

Later, he surmised that they could have been "sailed over."

When the flood hit the Nicholson home in Trailerville, Jerry had looked out the door. He noticed the other mobile homes begin to move out into the current like toy boats. Then their own trailer began to drift.

In the darkness Pat began to pray, "Lord Jesus, help us!" Then another mobile home crashed into them, ripping a five-foot gash in their fragile home.

That gash was a specific answer to their prayers. It was a way out! The Nicholsons began to struggle uith a sofa that had kept afloat in the rising water. Suddenly, it popped open and a mattress fell out.

As it did, Jerry heard a desperate cry from the water. He saw an arm and a head, apparently a woman in distress, but then she was gone.

Both of the Nicholsons' twin sons had taken swimming lessons. Tim got on the mattress, but Michael seemed to keep falling off. Finally, he just seized the foam mattress and held on. Pat and Jerry ended up alongside the mattress, one on each side.

At about that time, the bus came by. It didn't stop!

Once the Nicholsons neared the shore by what is called the old farm, Archie Smith hauled them out of the water. He was unclad, but they hardly even noticed. He ushered them through a barbed wire fence as he had helped others. Though Archie's back later showed a number of lacerations from where he had held up the wire fence, he didn't seem to notice the pain.

When Archie said, "All my family is gone," Jerry impulsively replied, "We'll be your family."

Jerry, Pat and the boys joined the group on the hillside, where they found sadness and serenity. There was a beautiful spirit in the group. There was love and compassion. And even humor.

Somebody said, "That's one way to get a pool in your back yard." The pool's water, they decided, was "sure dirty."

Archie Smith is a big man. Maybe 300 pounds. And everyone was cold. So they all huddled around him. "Come ahead," he invited, "fat is where it's at!" This from a shy man who thought his family had not survived.

But when the news came that Archie's family was safe, everyone rejoiced together.

The humor, though surprising, broke the tension and proved helpful. There is no mistaking Jerry and Pat Nicholson's seriousness. They believe God saved their family in a remarkable way and they are very grateful.

They believe the Scripture which says, "In every thing give thanks: for this is the will of God in Christ Jesus concerning you" (I Thessalonians 5:18). And they believe, like the rest of the people in Trailerville, that it should be practiced.

119

Archie Smith is another who believes it.

The water caught the Smiths on the floodplain fleeing toward Upper Trailerville. Archie saw they weren't going to make it, so he grabbed an exposed sewer pipe. When he saw a car and a trailer coming straight at him he abandoned the pipe.

He submerged, but when he came up he was inexplicably right beside a student's car. He grabbed the bumper and floated to safety—in the area of the old farm. Strangely, he remembers that the taillights were on.

When he got to a tree, the car stopped, so he seized the tree and the car promptly sank. Later, he scrambled to safety.

According to one report, twenty-seven trailers were swept away. The next day, in front of the few that remained, fragile flowers still danced in the breeze.

A note, in a young boy's handwriting, carried a request to be awakened the next day. It also said, "I love you, Mom."

A dog named Rufus made his own escape. Chained to his doghouse, he just pulled it up the incline.

Actually, in Trailerville there are many more accounts. All valid. Sometimes conflicting. But true. And powerful.

Mercifully, the waters moved on toward the concrete bridge over Route 17.

It held. Had it not, the toll would have mounted downstream. But because the bridge over Route 17 held, it became a secondary dam which sent a wall of water in reverse—a final stroke on the shattered remains of Trailerville.

...the effectual fervent prayer of a righteous man availeth much.

James 5:16

Chapter 18

We Swimmed and Swimmed

There is one more painful spot in Trailerville. Bill Ehrensberger's house, where only one person survived.

Bill had already been pulled off his feet and was gone.

When Eldon Elsberry stared at Ehrensberger's home and it began to float away, he was powerless to save Peggy Ehrensberger and her four children, even though his own ordeal was past.

But Tommy, one of the small boys, did survive. Robby, Kristen and Kenny were swept away with their mother.

All we know is that little Tommy was thrown into the water and that at one point he had hold of his older brother Robbie's hand.

"We swimmed and swimmed," says Tommy, pathetically.

Apparently, too, he climbed onto a log, only to be knocked off. Finally, Doug Veer, who had barely survived himself, pulled Tommy from the water.

While he waited for news about his family, Tommy was enveloped in the love and concern of one of the groups of survivors on the south shore of Toccoa Creek. They held him close and kept him warm.

When he asked about his mommy and daddy and wondered when they were coming for him, Tommy was told that his parents were missing and that no one knew where they were.

Tommy's stoic reply was, "That's okay."

Later he asked where Robbie was. But when told, "He's gone too," Tommy said, "Well, I thought so with all this water."

Tommy is managing. He's getting along. And once again he is wrapped in love, in St. Mary's, Pennsylvania. He is staying with Peggy's parents and Bill's parents live nearby.

Those dramatic last moments in a reeling, tilting home are unknown. The entrance that Peggy and her three children had into God's presence cannot be described till we see at last through the veil that clouds our vision from this world to the next.

The story we can't tell is what happened when the heavens opened. And surely they opened wide at 1:30 a.m. that dark November morning.

Within thirty seconds many of the Toccoa victims were already in. Stephen, who went in centuries before them, and whose words are recorded in the Bible, said as he was about to cross over, "Behold, I see the heavens opened, and the Son of Man standing at the right hand of God."

What those who entered from Toccoa Creek in flood saw we cannot tell. But Whom they saw we know. Some of us who read (and write) these lines find these witnesses and their words almost too powerful. Clearly, these people "sorrow not as others who have no hope."

What splendor and power would fill these pages if we could load this text with explosive details from the life that is "far better."

Impossible, of course. But mark it well. There is a heavenly side to this Toccoa story. Incredible. Intense. And

full of gripping detail.
    One day, most certainly, that story will be told.

> *...Eye hath not seen, nor ear heard, neither have entered into the heart of man, the things which God hath prepared for them that love him. But God hath revealed them unto us by his Spirit....*
>
> I Corinthians 2:9–10

IV.
Out of the Valley

Chapter 19

Twelve Helicopters on the Lawn

There was that sound.

Then that door.

Then that stretcher. And Ken Sanders, basketball coach of the Toccoa Falls Eagles, knew that death was there again.

Advised of the dam break by one of his players, Tom Penland, Sanders had descended from his home on higher ground to the campus at about 2:30 a.m.

He found no panic, only silence and a stench he had never smelled before. As Dean of Student Services, Sanders felt an immediate concern for Dave Eby and his family. (Eby served under him as Dean of Men.) For that reason, he moved toward the still-angry stream.

Then there was a cry across the creek where Eby's house once had stood.

"Dave?" said Sanders.

"Who is it?" the voice responded.

"Sanders. Is that you, Dave?"

It was. He and his family were all safe. Eby slid across the swollen creek on a fallen tree, and the men embraced and wept. Blood from a wound on Eby's arm seeped through Sanders' shirt.

But death was in the air. Someone had already seen a protruding human hand in the mud and chaos left in Forrest Hall. They hoped and prayed there wouldn't be more victims — but they were afraid there might be.

They headed for the hospital to see the injured and anxious. Death they left behind. Or so they thought.

Then there was a sound. The turning of a knob. The opening of a door. Then down the hall came a stretcher and death was there again. They had not left it at all. Sanders was asked to stay and help identify the dead. He stayed and did exactly that.

"I cried those fifteen hours," says Sanders, "seeing some I loved, some I knew, and others I would have loved if I had known them.

"The hurt came most when the hospital looked clear and then there was that sound, and then that door, and then that stretcher, and I knew death was there again.

"The room filled up again with friends and I cried, 'When will it end?'"

At 6 p.m., two were still missing, and darkness fell over the small Bible college. Sanders went home.

Stephens County Hospital, a solid brick building surrounded by spacious lawns, sits on land deeded to it by the college.

It is also close to the bridge across Toccoa Creek that allows Georgia's Route 17 to roll into the city of Toccoa. All the victims were swept in that direction and those that could headed directly for the hospital.

The scenes at the hospital will never be forgotten. Some victims arrived who were unrecognizable because of injuries. There was a father who followed stretcher after stretcher into the makeshift morgue — till he found what he hoped he wouldn't find.

Ruth Payton, a registered nurse, helped Sanders and Dave Wilson, a student, identify many of the bodies. She knew most of the victims personally and prayed before

127

the first identification, "Lord, help me to remember that whoever this is, I am just identifying the shell that once housed the person, and that he or she is already rejoicing in Your presence."

The bodies of the families were kept together in the morgue.

About 10 a.m. the Governor of Georgia, George Busbee, and his entourage arrived. The army was coming in too. Later, Senator Herman E. Talmadge and Congressman Ed Jenkins also visited the campus. Busbee was present when the bodies of three children were pulled from the carnage in the creek.

A decision had been made earlier to take the bodies to the two funeral homes in Toccoa where they were embalmed. Some of the bodies were later shifted between the mortuaries according to family preference.

After the bodies were embalmed they were taken to a temporary morgue in an elementary school gymnasium in Toccoa.

Both the Governor and the Commander of the National Guard operated out of the hospital complex. Governor Busbee told Hospital Administrator J.W. Warren that U.S. President Jimmy Carter's wife Rosalynn was coming. Shortly after, her secret service detail arrived.

Warren, the Governor and Kenn Opperman all greeted Mrs. Carter. She had flown to the Anderson Air Force Base in South Carolina and then had been driven over the state line into Georgia.

The media also made the hospital a focal point and at one time there were a dozen helicopters on the manicured lawn.

America's gracious First Lady visited Jeff Metzger first. Jeff knew that his wife and son were dead. But despite his pain, he was cheerful.

Mrs. Carter also visited others who had lost loved ones or been injured. She found a similar phenomenon.

At one point she is reported to have said, "I came down here to help and give what little comfort I could to the victims and their families, but I feel I'm taking away more than I'm able to give."

There were some unusual mental health aspects to the disaster. At one time. Warren believes, there were between 150 and 200 students jammed into the halls and doorways. But generally, there were no "hangups" and very little hysteria. There was, instead, a general acceptance of God's will.

The bodies of children brought the tears. The second body brought in was that of a child. Pathetically, one little child's hand held a stick.

Warren, a hospital administrator of twenty-nine years experience, admits that the sight of that little victim broke him up. He has a daughter about that age.

The events at the hospital also had a spiritual impact on Warren. In his own words he says, "My faith was deepened." He had been a churchman all his life, but he gained "inside knowledge" during the disaster's aftermath.

Dr. John Payton, formerly an Alliance missionary to Thailand, greeted the ambulance driver when he arrived at Stephens County Hospital Monday morning. "Theo, thank you for being so good to our people," he said.

The driver replied, "Dr. Payton, those people were easy to care for. We've had more hysteria at a two-car collision than we had all that night and the next morning!"

Another hospital official, when questioned, told Payton, "I am sure there must have been some hysteria among the younger students last night, but we saw none."

When Payton checked the records for the night, only one shot of valium had been given and only one shot of narcotics had been administered to the victims.

The one girl who did become hysterical had been on a date, just the night before, with one of the fellows who had drowned in the dorm.

The local mental health officials, understanding the unique nature of the Toccoa Falls community, knew that extensive mental health counseling would not be needed.

Not everyone understood, though. At one point, a Washington-based mental health official insisted on searching out victims who needed psychological counseling. And she wasn't too happy to discover that there was little need for her services.

When the message was relayed to Dr. Payton he responded, "We have about five hundred of the most accommodating people in the world out here at the college. If she needs victims, I'm sure we can find some volunteers!

A thoughtful man, John Payton cites two main reasons for the scarcity of severe mental stress.

"Among evangelical people there is the practice of child dedication," he says. "Parents formally present their children to God for Him to enact His will in their lives. Later—if the children should die—parents find death easier to accept. They have already given those children to God.

"Secondly, in the case of Toccoa Falls, Dr. Jerry Sproull had exerted a considerable influence upon the young couples' marital relationships. He had encouraged the couples to always deal with conflicts immediately, to resolve all strife and thus to fulfill the biblical injunction, 'Let not the sun go down upon thy wrath.'"

The wounded did weep and bleed, of course, the morning of November 6 in the Stephens County Hospital. But in that, too, there was strength, serenity and power.

W.A. "Bill" Warren summed it up for everyone when he told an interviewer, "If you can capture what we saw here at the hospital, people will be knocking down the doors to get into this college."

But for Coach Sanders, the trauma was not over.

On December 3, nearly a month after the disaster, a student came to Sanders at the college Christmas banquet. Again it was a basketball player, John Penland. He was concerned that one of the students had not arrived. Soon it was learned that there had been an accident. Two cars had collided, and Mrs. Bertha Walls, the youthful wife of a young man who had replaced Betty Jean Woerner as Dr. Opperman's secretary, had been injured.

After he had learned of the accident, Dr. Alvin Moser, college Vice President, proceeded to deliver his address. Its theme: though Bethlehem was small and obscure, God chose to make it widely known as a place of joy and blessing where multitudes return figuratively every Christmas to find hope and peace. In a similar fashion, God has been pleased to make obscure Toccoa falls a symbol of this same hope and peace to the world.

Then once more Ken Sanders and Alvin Moser headed for the Stephens County Hospital. And one last time they stared death in the face. Bertha Walls was gone too. Death had visited the President's office yet another time.

Finally, the hand of death was stayed at Toccoa Falls. And life, warm as the breath of spring, today blows across the campus. The ordeal by accidents and flood is past. A new day has dawned.

It doesn't matter anymore that Toccoa Falls had been in difficult financial straits before the flood. Nor is any credence given now to the rumors that said the college could never survive. The old school that has served so well for so long is not going to die after all.

She really is the "planting of the Lord."

> *Remember ye not the former things, neither consider the things of old. Behold, I will do a new thing; now it shall spring forth; shall ye not know it?*
>
> Isaiah 43:18,19a

Chapter 20

Stability Comes Through

"Dad, the whole campus is washing away!"

Dr. Kenn Opperman, President of Toccoa Falls College, was wakened in Florida by a call from his son Bob, at home on campus in Toccoa Falls.

Over the phone, as an awesome background to his son's voice, Opperman heard the snapping of timbers and the roar of water four-tenths of a mile away. But he could not be there to help.

Dr. Alvin Moser, Vice President of the college, was the highest administrative official on the grounds. It fell to him to gather the students to lead and encourage them.

The college was in shock, and stability and calmness were needed.

Moser prayed for daylight.

When the water first began to rampage through the lower campus, the several hundred students and other college personnel were uncertain as to where it would go, or how much water would come. The dormitories emptied and students milled around. Vehicles seeking to negotiate the campus roads found their way blocked by anxious students gathered in little clusters. The students prayed. There was weeping and great distress. And a thousand times the questions were asked, "Where is...?" "Have you seen...?"

Ironically, there were others who slept through the whole night. They had not known until the next morning that the dam had broken.

In a service conducted by Dr. Moser at 10 a.m. in the chapel the students offered testimonies and requested prayers. Songs were chosen and sung. The stability of the Christian faith was coming through.

> Stayed upon Jehovah,
> Hearts are fully blessed,
> Finding as He promised,
> Perfect peace and rest.

The flow of rumors ebbed and rose.

By that time the news media had arrived and the students' reactions were being chronicled for distribution across America and the world.

Apparently, too, the newsmen were not prepared for the openness and guilelessness of the students. Some reporters, who followed disasters regularly, had never encountered reactions like they found at Toccoa Falls.

College personnel were also vigorous in their witness. They talked simply about the disaster and about their faith in Jesus Christ. When one student asked Mrs. May Trucano, editor of the college quarterly **TFC Today,** when the reporters would leave, Mrs. Trucano's answer was precise—"They will all go when God is through with them."

The attitude toward death, the spontaneity and the joy among the students of the college were striking to the world outside, but Vice President Moser believed the disaster only illuminated what was present in their lives all the time. Undoubtedly, he is right.

It must be said, too, that not all was joy and peace. There was heaviness, and hurt and tears.

Some of the students wondered if the college would be able to survive. On Sunday morning a meeting was convened in the chapel and a number of announcements were made.

One of the first was made by Mr. J. J. Fowler of Atlanta, chairman of the Board of Trustees. "This is not the end of Toccoa Falls College," he said. "It's just a new beginning."

Immediately there was a thundering emotional response from the audience. Sustained clapping and applause. A standing ovation. Below the open concern for loved ones and friends, apparently there had been a deep anxiety about the future of the college itself.

Photographers and reporters in the back of the auditorium could hardly believe what they were witnessing. The father of one of the students, overcome by emotion, slipped out of the hall to pray. He wanted desperately to get right with God.

By five o'clock that afternoon the campus dormitories had been vacated. Most students had returned to their homes. Others had gone to stay at the Southern Baptist conference center at nearby Lake Louise, Georgia.

In the days that followed, the Civil Defense under General Billy Jones carried out a massive cleanup that quickly erased most of the signs of the flood.

The offers of help came from all over America. Sometimes there were so many offers that there was no way to cope with them.

Inmates from a nearby correctional institution, some of whom had been led to Christ by students, volunteered their services. At one point, down near the floodplain, the inmates offered to help load some cattle into a truck. One of the animals bolted and more than twenty inmates, in hot pursuit, followed it over the hill.

The dumbfounded guard was heard to say, "This could be trouble!"

But they all came back.

And later they caught the animal too.

The day of the Memorial Service, held November 16 at 8 a.m. in the Student Center, was a memorable one. Heavy showers preceded the gathering and a magnificent double rainbow hovered over the campus. Many saw it as a symbol of beauty and hope for broken hearts. It also seemed to say, "Toccoa Falls College will never be destroyed by flood again."

The service was simple and full of deep dignity.

There were expressions of thanks, majestic music and a healing message of optimism by President Opperman.

He said, "For those who can't quite comprehend the great victory, the smiling faces, the victorious statements — for those who say, 'How can this possibly be?' — the answer is found in the nature of the family that is drawn together in Toccoa Falls College. It's made up of young men and women from around the world who need His strength, who have had an encounter with Him, who know that they don't have to return but that they do have to go — and so, to the best of their ability, they're training to go.

"Many worldwide organizations have sent us letters and communications offering us aid — and for all such help I am indeed grateful — but some of the deepest feelings of my heart have been for people who lost everything... house, accumulations, furniture...everything. In two cases, married students who had nothing left on this earth went to the bank and withdrew every penny of their savings. Instead of saying, 'I've got to survive,' they came to me with their checks and said, 'Toccoa Falls must survive!'

"How can you explain that kind of dedication?

"Sixty-four years ago when Dr. R.A. Forrest sat on a stone with a stick in his hand, almost absentmindedly working his way through the ashes of what had once been Haddock Inn, the first home of Toccoa Falls College, he sat there with tears running down his cheeks. Then he heard a voice speak to him from heaven. And as the Lord drew near to him, He seemed to say, 'Weep not, for thou

135

shalt have beauty for ashes. Dost thou not know? This school is the planting of the Lord. It shall continue.'

"Our young people recognize that life is filled with trouble," he continued. "Yet I find among these Christian youth a radiance, an acceptance of God's will, an awareness that God has done something very beautiful for them.

"Many have asked us, 'What accounts for the radiance of the people here?' I have only one answer, an answer that a person who isn't a Christian can't accept. It's found in the Bible: 'Christ in you, the hope of glory.' For a Christian the most important thing is a relationship with Christ, so that whether we live or die we glorify the Lord Jesus.

"There's something unusual here at Toccoa. It's found in young people who are already committed to the fact that life is meant to be lived every day and death is not something to be dreaded. We mourn the thirty-nine who are no longer with us, yet we know they are with their Master.

"Toccoa Falls College is going to continue because we believe we need to train men and women to preach the gospel of Jesus Christ until the King Himself returns. We intend to do that."[1]

...Christ in you, the hope of glory.

Colossians 1:27

---

1. The Alliance Witness, December 28, 1977.

Chapter 21

Disaster with a Difference

The coordinator for all agencies of the U.S. government in the Toccoa Falls disaster was hunched behind the chair in front of him — praying and crying.

At Toccoa he had found a disaster with a difference — and even in the crowd of 1,500 that had gathered in the Student Center at 8 a.m. November 16 for the Memorial Service, he had been unable to escape its personal implications.

Tom Credle, 43, from Atlanta, is not a man easily moved to tears. But at Toccoa Falls, the disaster coordinator for the U.S. government, a veteran of nineteen previous disasters, a pragmatic man with military background, found himself strangely overwhelmed with emotional and spiritual concern.

He had sought out the chair at the rear of the auditorium because as the "head fed," the top governmental official on the scene, Credle had a feeling that Dr. Opperman might call upon him for a few words. And he considered himself in no shape for words of any kind.

Kenn Opperman said to him after the service, "Where

were you? I was going to call on you." Credle's "feeling" had been well-founded!

Tom Credle had heard about the flood at Toccoa Falls the morning of November 6. He was on the scene early in the day, and his first impressions were not unusual at all. It was another disaster, with the usual debris and suffering. But as one of the ten federal disaster coordinators in the country, he was responsible to set up an office and to coordinate work in the area until the cleanup and repairs had been accomplished. He returned to Atlanta to prepare a summary of his findings.

When a decision had to be made at Credle's office about whom to send to Toccoa, Tom Credle chose to go himself.

It was a casual decision — that ultimately proved to be momentous.

Once on the campus, Credle soon observed that his relationship with God seemed anemic when compared with that of the students and faculty at Toccoa.

Also, uncomfortably, a question began to hammer his mind — "Why did this happen to Toccoa Falls?"

Never, in any other disaster, had he become personally involved. Never had he had any questions. But now there was an overpowering concern. So overpowering, in fact, that it drove him to his knees in his Toccoa motel.

Why had this happened to Toccoa Falls? Why? Why? "Maybe, Lord, You can tell me why," he prayed.

Part of that answer came soon — back on the campus. Credle didn't realize that capable graduates from Toccoa Falls were actively engaged in full-time Christian service in fifty-six countries of the world and in all fifty states. No matter. Credle's conclusion was clear. "Toccoa Falls and the mission that it has must continue."

Tom Credle also knew that the college had been in a desperate situation and flat on its back financially even before the flood. And he realized that the dramatic events had suddenly pushed Toccoa Falls into national focus. At the price of thirty-nine lives, it had been thrust

138

into the national consciousness. Undoubtedly, Toccoa
Falls had become, overnight, the best-known small college in the United States. The response of the U.S. public
and the U.S. government had something very vital to do
with the college's survival.

Tom Credle's personal spiritual odyssey was far from
over. During subsequent days on campus he frequently
visited Dr. Opperman's office, and not just to iron out
details or procedures. He went, he says, to draw strength.
To talk and to pray.

In private moments he noticed, too, that his knees were
more limber than they had been in the past. It was easier
to pray. More comfortable in God's presence. His questions, however, were only partly answered.

But as Jimmy Carter's man on the scene drove toward
the Student Center for that early-morning Memorial Service, the second part of his private puzzle fell into place.

A massive, richly colored rainbow arched upward into
the sky over the campus. He was awestruck — so much so
that he asked others if they had seen what he had seen.
They had. And its message was clear.

Toccoa Falls College would continue. God would see to
that. God knew what He was doing at Toccoa.

And now Credle knew it too.

> *I do set my bow in the cloud, and it shall be for a
> token of a covenant between me and the earth. And
> I will remember my covenant, which is between me
> and you and every living creature of all flesh; and
> the waters shall no more become a flood to destroy
> all flesh.*
>
> Genesis 9:13,15

Chapter 22

Why This Response?

A visiting reporter on the campus of Toccoa Falls College asked Professor Gerald E. McGraw how he was ever going to vindicate God in the eyes of his students after such a great disaster.

McGraw is a tall theologian who moves and speaks very deliberately. His reply in this case was very careful.

"The question has never even come up," he said honestly.

To a reporter unschooled in biblical ideas, Dr. McGraw's reply was stunning. His reply, we are sure, will also puzzle some who read this book. But then, his response is just one more of the highly unusual reactions chronicled between these covers.

Frankly, there is no possibility of understanding a book like this without first understanding the responses that come from minds and hearts steeped in the Bible and determined to obey its teachings.

Toccoa Falls College is a theological institution and there was clearly a theological response to the flood at Toccoa Falls. The Scriptures teach that God is sovereign, above all, and ultimately in control. Perhaps for that reason the text summoned more often by these people in

140

the midst of their disaster was this one: "And we know that all things work together for good to them that love God, to them who are the called according to his purpose."[1]

This perspective is the real explanation for the relative serenity that prevailed on the ravaged campus at Toccoa Falls.

Why does President Opperman reach back into history to use such phrases as "beauty for ashes" and "the planting of the Lord"?[2] He believes, as does the whole community, that the Bible is without error, inspired, alive, eternal, and completely trustworthy. That's why words written by the prophet Isaiah or by King David thousands of years ago are considered helpful and encouraging today. "All Scripture," Opperman would say with Saint Paul, "is given by inspiration of God, and is profitable for doctrine, for reproof, for correction, for instruction in righteousness."[3]

Reporters did not expect to find verifying evidence of the truth of Christianity on the Toccoa Falls campus. But expected or not, it was there.

Actually, there was no escaping it. The whole college community demonstrated a deep, personal knowledge of Jesus Christ. That's why Dr. Alvin Moser affirms in these pages that the disaster only illuminated what was present all the time in the lives of the students. That is why Dr. John Payton told us, "Now I really know that what I've been telling others about Jesus is true."

Surprisingly, the flood supplied hard evidence, and plenty of it, that Jesus Christ is alive.

The thanksgiving in difficult straits is also nearly impossible to understand unless one is governed by the biblical concept, "In every thing give thanks."[4]

---

1. Romans 8:28
2. Isaiah 61:3
3. II Timothy 3:16
4. I Thessalonians 5:18

That's a high order. You may have been amazed that a little boy locked in an air bubble in a truck cab actually did it. Perhaps you were puzzled when a beautiful young widow, just bereaved, found strength to do it. Clearly, the Bible drastically alters human behavior.

Ordinarily, in disasters such as the one at Toccoa Falls, there are all kinds of threats, recriminations and words of anger. They just were not heard at Toccoa. Somehow, the reactions of these people of the Book were different.

The way they faced death was different too. There was strength, serenity, calmness. It seems that believers who have invited Jesus Christ to be their Lord and Savior somehow know, intuitively and deeply, that they have eternal life. Death is neither welcomed nor sought by them, of course. But the dread is gone. Death, for them, has lost its sting. "O death, where is thy sting? O grave, where is thy victory?"[5] With the Apostle Paul they say, "...absent from the body...present with the Lord."[6] There is a whole philosophy of death and dying in these pages that can only be explained by the Scriptures.

At Toccoa Falls these biblical ideas were like the rails that carry the trains through storms and darkness. They provided stability and direction. Death simply did not threaten these people the way it threatens casual Christians or unbelieving humanists. Christians have always died well, and they always will.

In the next chapter you will read a letter from man who was watching a television newscast. And we'll tell you this much now, as he watches the reaction at Toccoa Falls, he is converted to faith in Jesus Christ. The same thing happens to him that happened to Bill Anderson's brother pages earlier. (Remember, he didn't come to Toccoa for the funeral, he came to get what Bill had.) The same thing has been happening repeatedly as people have responded to the events at Toccoa Falls, and it is only

---

5. I Corinthians 15:55
6. II Corinthians 5:8

partially recorded here.

What is the meaning of all of this? Again, it is the application of biblical ideas to life. The man who watched the newscast in which flood survivors told their stories apparently knew some important biblical teachings.

Somewhere he had learned that Jesus Christ died for everyone. "For God so loved the world, that he gave his only begotten Son, that whosoever believeth in Him should not perish, but have everlasting life."[7]

Further, he also obviously accepted the Bible's verdict as applying to himself. "For all have sinned, and come short of the glory of God."[8] Apparently, he felt a personal need to turn from his sins.

Somewhere he had also heard the biblical concept of personal responsibility. "But as many as received him, to them gave he power to become the sons of God, even to them that believe on his name."[9] He knew he had to do something. And the witnesses on the telecast pushed him over the line of decision.

Finally, he obviously acted upon what he had known previously, and perhaps what he had felt through the telecast. He called upon the Lord. And it is the Book which says, "Whosoever shall call upon the name of the Lord shall be saved."[10] In biblical terminology, he had been saved, born again.

When he wrote the college, whether he knew it or not, he was acting upon yet another biblical principle. "For with the heart man believeth unto righteousness; and with the mouth confession is made unto salvation."[11] Real believers always make it known.

What happened to that viewer will happen to some of

---

7. John 3:16
8. Romans 3:23
9. John 1:12
10. Romans 10:13
11. Romans 10:10

our readers too. We know that. The accounts in these pages are like bare electric wires. When you seize them, you feel the power.

We do not object, of course. We're pleased. But we think you should know how it happens and why. God the Holy Spirit takes the biblical ideas that run like a spine through the dramatic story we have sought to tell here. And He makes you think subjectively, about yourself. To begin praying when you're reading this book is as natural as can be.

Perhaps it has already begun. In any case, you need not stop calling upon the Lord until you are absolutely sure about Jesus Christ, until you really know that you have eternal life, until you really understand the words of John, the Apostle: "These things have I written unto you that believe on the name of the Son of God; that ye may know that ye have eternal life...."[12]

Later, you will look back and say, "Now I can really understand what happened at Toccoa Falls. I really can."

---

12. I John 5:13

Heaven and earth shall pass away, but my words shall not pass away.

Mark 13:31

Chapter 23

A Robin on the Roof

Before the sun ever dawned that dark morning in November, a robin began to sing heartily on the upper reaches of LeTourneau Hall, a women's dormitory on campus. What its loss had been, no one knew. But it had a song. A blind student at Toccoa, who could not be overpowered by the sight of the devastation all around, was the one to hear the tiny bird.

But President Opperman was not there to hear it. He was hundreds of miles away in Florida and pressing hard to get back to Toccoa.

On July 4, 1976, Opperman, 52, had suffered a heart attack. Seven more attacks followed, culminating on July 13. For the following ten weeks, Kenn Opperman, the former missionary to Peru, hung on a thread between life and death.

On September 27, 1976, open-heart surgery was performed and a week later Opperman walked out of the hospital.

One of the reasons for the nearly fatal attack, he now believes, is that he wasn't entirely trusting the Lord whom he serves. The financial needs of the college and his obligation to meet a payroll month after month without resources seemed an insuperable burden.

Still he believed God would intervene in the life of Toccoa Falls College.

The Scripture to which he had clung was Psalm 66:12, "We went through the fire and through water, but thou broughtest us out into a wealthy place." Not until after November 6 did he really observe that water was mentioned in the text.

The weekend that the dam burst, Opperman had gone to Florida to participate in the funeral of Richele Hornbeek. Richele was a popular sophomore who had been critically injured on the way back to Toccoa Falls after a long weekend at home.

Her death, when it came, had a remarkable effect on her home church, the First Alliance of Orlando. Many of her friends were on her prayer list, which was made public at the funeral, and they were deeply moved when they discovered that they had been the subjects of Richele's prayers. The fatal accident also profoundly touched Toccoa Falls College while she lingered in a coma after the accident. Her death caused the whole college family to think about eternity very seriously. Just on the eve of November 6.

At 1:45 a.m., while the flood was still in progress, Bob Opperman's call reached his father in Florida.

He had become aware of the commotion and had reached the lower campus area in time to see the aftermath of the big wave go through. Though young Opperman thought most of the college was washing away, later it became clear that only about one-third of the campus had been affected.

Vice President Moser also called Opperman in Orlando. The college President sought assurance that order would be maintained and that no students would attempt heroics.

The 9:30 a.m. flight from Orlando to Atlanta was booked solid, but when the nature of Dr. Opperman's

146

mission became known the airline bumped a member of its staff to accommodate him.

Mr. J. J. Fowler, Atlanta businessman and chairman of the Board of Trustees, met Opperman in Atlanta and together they drove northeast toward Toccoa. At the outset of the journey their car radio reported eleven dead. By the time they arrived at Toccoa, the known casualties had reached thirty-three.

Because of the continuing water flow over the bridge at Route 17, the bridge was not passable and a policeman forbade Opperman to cross to Stephens County Hospital and the stricken campus on the other side.

"But I'm the college President," Opperman objected.

"You're the twelfth college President who's tried to cross here," he was told.

Dr. Opperman came on across the bridge anyway. And as he moved through the shallow water toward his devastated college, in the movement of debris, a child's body surfaced. It was that of a little girl. Tenderly, Kenn Opperman picked up the tiny form and handed it to the wondering policeman.

Opperman headed for the hospital, where he verified the identifications of some of the bodies. There, too, he learned that his secretary, Betty Jean Woerner, and her daughter had also died.

General Billy Jones of the U.S. Armed Forces, Civil Defense, directed procedures for handling the tragedy. Some looting had already taken place and more was feared.

The looting, though not widespread, was irritating. And it was in stark contrast to the magnificent outpouring of sympathy and help that immediately began to arrive.

The governmental assistance was instantaneous and generous.

There was an overwhelming and deep response from the citizens of Toccoa and its environs.

Georgia's Governor George Busbee remained on the campus Sunday afternoon. First Lady Rosalynn Carter, visibly moved by what she had seen, offered her help and promised to encourage the President to declare Toccoa Falls a Federal Disaster Area.

President Carter had heard the news in the First Baptist Church in Washington, and had decided during the service to ask his wife to go to Georgia. Mrs. Carter's presence on the Toccoa Falls campus dramatically focused the attention of the world on the disaster and a battered but determined college family.

The day after the flood, the World Relief Commission of the National Association of Evangelicals presented a check for $50,000.00.

The Christian and Missionary Alliance pledged $25,000.00 and, to date, has sent $42,000.00 to Toccoa.

Dr. Billy Graham dispatched an associate, Dr. Grady Wilson, to personally deliver $25,000.00 and he also sent this telegram:

> Just heard of the tragedy and glory of the flood on my way to our next crusade in the Philippines. While shocked and grieving for those that lost loved ones yet thrilled at the glorious witness for Christ on the front pages of the newspapers. Have given instructions for our organization to do all in its power to be of help and am certain that God has a purpose and that His grace will be more than sufficient.

Pope Paul sent personal condolences by way of Archbishop Donellan of Atlanta. Paul is the pontiff with whom Opperman had argued years ago about the persecution of protestants in South America, but in subsequent conversations their relationship had mellowed and, remarkably, in eight out of nine audiences over a span of years, Pope Paul had questioned and probed Opperman, "How do you know you have eternal life?" Repeatedly, Dr. Opperman had cited biblical passages and affirmed that for those reasons, he knew he had eternal life.

And the aged Pope had not forgotten.
The Vatican's telegram read:

> Holy Father grieved to learn of disaster in
> Toccoa causing such great loss of life and suf-
> fering. He would ask you to convey to bereaved
> families the expression of his sincere sympathy
> and his prayerful solidarity with all afflicted by
> the calamity.

There were other outpourings of grief. At least one hundred telegrams and letters from college presidents and
student groups came in.

Several colleges offered to integrate Toccoa students
into their own programs. A number of other colleges
fasted and sent the money they would have spent on
food.

A Mormon college, which had also gone through a dam
break and a flood resulting in multiple deaths, sent expressions of concern and a substantial gift.

A Methodist college staged a variety show—and sent
the proceeds.

Broadcasters also did their part. Dr. Robert H. Schuller
invited Dr. Opperman to speak on his nationwide television program. He then invited his audience to send help
to Toccoa Falls College. And Schuller's loyal audience
has been doing exactly that.

Pat Robertson, genial host of **The 700 Club** and the
leader of the Christian Broadcasting Network, also gave
Opperman an opportunity to tell the Toccoa story. Rob-
ertson himself was deeply moved by what he heard.
Later, his association forwarded $50,000.00 to Toccoa
Falls College.

The flow of monetary gifts came from all over the world
and has mounted until this day to over a million dollars.

Every day the mail brought additional expressions of
love and concern.

From Arizona, for example, came this letter from an Indian Mission:

> The people here at Pinin Navajo Gospel Church wanted you to know that they care about you and have been praying for you...
> On hearing of the disaster at the Bible School, they wanted to help, so they took up this offering at a prayer meeting. One lady had no money so she put in her necklace. Maybe you can sell it and use the money....

A 104-year-old American woman living in England commended Toccoa on "the spunk" with which they were facing the tragedy and sent $2,000.00
A man from Pennsylvania wrote:

> I was very sorry to hear about your disaster with the flooding. In a way I praise God for it though. Let me explain why. On the news I saw how your people at the college were still praising and loving God. I gave my life to Jesus during that newscast, and I now have Jesus in my heart....

Another man wrote, "My wife and I were converted yesterday. I have just realized that God has been blessing me all my life and I have a lot of catching up to do." He included a gift for the college.
A pastor from Florida wrote that he had sought a word from the Lord concerning what had taken place. The answer he received was Psalm 46:5: "God is in the midst of her; she shall not be moved: God shall help her, and that right early."
Many of the letters contained checks to help restore the beautiful campus and meet other pressing needs.

A pastor in South Carolina, for example, approached several manufacturers for help and support. In all, he raised about $8,000.00 plus a quantity of goods — including twelve cases of diapers — Pampers!

One of the letters that President Opperman remembers most fondly came at a time when he especially needed a lift. This letter, from a Roman Catholic priest, provided encouragement and a laugh too: "Time will go by and people will have forgotten something. So here's $100 to buy grass seed."

One woman said, "Here's the widow's mite. This is all I've got."

Yet another woman sent the money she had saved for her husband's Christmas gift. She explained, "My husband wanted to help Toccoa Falls more than anything else in his life. When you accept this gift you will give him the desires of his heart."

We could continue — for the letters, telegrams, and offers of help go on and on. And so do the love and concern expressed toward Toccoa Falls College.

But needs remain. Total damage to the campus and its residents, along with restoration and recovery costs, has been estimated at more than six million dollars. The first $450,000.00 that was received includes all insurance claims and many gifts, and has been dispersed to students and others who suffered personal loss.

Only when that need was met have funds been applied to the recovery of the college.

Federal disaster funds designated for the college total two million dollars, but beyond that there remains a great need.

Opperman does not follow a policy any different from Presidents Bandy and Forrest before him. The college will simply trust in God.

As has always happened in the past, they believe here at Toccoa Falls College that God will provide. He will moti-

vate people to give gifts and God's work will go on.
They know it.

But my God shall supply all your need according to his riches in glory by Christ Jesus.

Philippians 4:19

Chapter 24

And Since the Flood

It is thirteen years since the flood.

Direct evidences of the damage are no more. Instead, on every side there are signs of vigor and life. The Falls themselves are strikingly beautiful and continue to attract thousands of visitors each year. There is a new 32,000 square foot gymnatorium and a new wing on the Forrest Hall men's residence. A music building and student union complex enhance the updated campus.

The town of Toccoa and its citizenry have also rallied to contribute to such campus projects at the Frankeline and W.C. Clary Science Building.

Under the leadership of Dr. Paul Alford, president since 1978, the student body has increased from 480 to 840. The budget of the College has gone from $1.8 million before the flood to $6.6 million now (1990). For thirteen years Toccoa Falls College has finished in the black and for the last two years has been debt-free, no mean feat given contemporary college statistics.

And what of the people who survived the tragedy?

Mrs. Pat Sproull, whom you met in chapter 8, has found her way to personal healing. She was among those who suffered the most devastating blows—her husband and three little girls all perished in the flood. Pat is now involved in a medical ministry in Guatemala.

And then there's missionary Ray Woerner. You will remember him as the man a few chapters back who saw his trunk floating by the bridge. Though he lost his wife and daughter, he felt compelled to stay on at the college for a while.

Just prior to the flood, Paul Carpenter of Corning, NY had donated an extensive array of photographic equipment to the college. Its loss meant an insurance payout of $65,000. That money, under the careful stewardship of Mr. Carpenter, funded the campus radio station WRAF, named after the founder of the college, Dr. R.A. Forrest. For 10 years, WRAF has been broadcasting at 100,000 watts FM to four southern states. A second station, WCOP (W Christ Oriented Programming), 5,000 watts AM, is now functioning at Warner Robins, Georgia.

Ray Woerner learned a great deal as he watched WRAF being launched on the Toccoa campus. When he headed back overseas, he carried his new awareness to Chile, South America, where he spearheaded Radio Esperanza (Radio Hope), the first evangelical radio station in that nation. A whole new chapter of missionary radio emerged from the flood at Toccoa Falls.

Dam Break in Georgia, the book in your hand, has been sold and given away in large numbers. More than 100,000 copies were sold immediately after the flood. In the years since, another 100,000 copies have been distributed, 20,000 of them inside the prisons of America.

Many dramatic stories have been reported as a result of people reading the book.

Not long after the flood, a Christian truck driver pulled into a truck-stop restaurant. Seeing a depressed-looking man sitting in a corner, he approached him. Could they talk, he wanted to know. The answer was a sharp rebuff. However, the impression persisted until finally the truck driver approached the man again.

"Listen, I am a Christian and I am supposed to talk to you." It turned out that the man was a new believer, a salesman. He had just phoned home, only to hear an unexpected voice on the other end of the line informing him that his wife and two small children had been killed in an automobile accident. He had been sitting there trying to summon the courage to go home and face his own appalling personal disaster.

The truck driver hurried to his rig for a copy of **Dam Break** and gave it to the distraught husband and father. Plunging in immediately, the devastated man read it all the way through. And early in the morning he headed for the truck driver's motel room. He was ready now, he said, to face the heartbreak at home. Strength and courage had come to one more person through the compelling accounts of the survivors of the flood at Toccoa Falls.

The saga continues, through the life of a college that has experienced both fire and flood — and a great deal of pain. The ministry also continues through the lives of those who survived the waters to emerge into an enlarging morning of joy.

Through it all, the Scripture that has been so meaningful in the life of the college still stands: "We went through fire and water, but you brought us to a place of abundance" (Psalm 66:12).

Epilogue

Writing **Dam Break In Georgia** has been a moving experience. The testimony which we have heard has been, if anything, more impressive than what we have managed to capture in print.

We realize that this story has hundreds of versions—all valid, though not necessarily in agreement. We have limited ourselves almost exclusively to interviews with the bereaved families, but there have been some exceptions. We have allowed some contradictions in the text because various witnesses have remembered things differently.

Much more, very much more, could be said about people and organizations who were involved in the disaster or responded to it afterward. We trust for their understanding in view of these omissions.

Dr. James Grant, Director of Public Affairs at Toccoa Falls College, was especially helpful. His overview of the story as a whole proved exceptionally valuable in the rapid production of this book. We are deeply grateful.

A number of reactions are possible, even likely, from a reading of this text. Some will discover our Lord and Savior too. We hope for this above all.

Others will discover a whole new philosophy for coping

with death and bereavement. They will discover, too, that this philosophy is not new, but is as old as the Scriptures.

Another result will appear as well. Many will want to help. We know that many will send gifts to Toccoa Falls. Not because the college has contrived or asked. But because our Lord works that way.

* Toccoa Falls College
Toccoa Falls, Georgia 30598

In Memoriam

Richele Marie Hornbeek

Karen Anderson
Joseph Anderson
Rebecca Anderson
Gerald Brittin
William L. Ehrensberger
Peggy Ann Ehrensberger
Robert Ehrensberger
Kristen Ehrensberger
Kenny Ehrensberger
David Fledderjohann
Mary Jo Ginther
Brenda Ginther
Rhonda Ginther
Nancy Ginther
Tracy Ginther
Cary E. Hanna
Tiep Harner
Robbie Harner
Christopher Kemp
Cassandra Metzger
Dirksen Metzger

Ruth Moore
Jeremiah Moore
Eduard E. Pepsny
Carol Pepsny
Paul Pepsny
Bonnie Pepsny
Eloise J. Pinney
Monroe J. Rupp
Jerry Sproull
Melissa Sproull
Jocelyn Sproull
Joanna Sproull
Richard J. Swires
Jaimée Veer
Mary N. Williams
Paul I. Williams
Betty Jean Woerner
Deborah Woerner

Bertha Walls

About the Authors

K. Neill Foster is the author of five books. This was his fourth. Eric Mills was Book Editor with Horizon House Publishers.

CREDITS:

Cover illustrations: Karl Foster. Maps: Brenda Rickey. Photos Inside: Paul M. Brown, Jerome McClendon, Joanne Schneider, Toccoa Falls College. Written materials: Bea Rupp, Ken Sanders.

For additional copies of

Dam Break in Georgia

contact your local Christian bookstore or call Christian Publications.
1-800-233-4443